RadioScience Observing

Observing

Volume Two

By Joseph J. Carr

RadioScience
Observing

Volume Two

By Joseph J. Carr

PROMPT© Publications is an imprint of Howard W. Sams & Company, A Bell Atlantic Company, 2647 Waterfront Parkway, E. Dr., Indianapolis, IN 46214-2041.

International Standard Book Number: 0-7906-1172-4
Library of Congress Catalog Card Number: 98-68108

Acquisitions Editor: Loretta Yates
Editor: Pat Brady
Assistant Editor: J.B. Hall
Typesetting: Pat Brady
Cover Design: Christy Pierce
Graphics Conversion: Terry Varvel
Illustrations and Other Materials: Courtesy of the Author

Trademark Acknowledgments:
All product illustrations, product names and logos are trademarks of their respective manufacturers. All terms in this book that are known or suspected to be trademarks or services have been appropriately capitalized. PROMPT® Publications, Howard W. Sams & Company, and Bell Atlantic cannot attest to the accuracy of this information. Use of an illustration, term or logo in this book should not be regarded as affecting the validity of any trademark or service mark.

Dedication

If the Lord decrees that only ten names can be written in the
<u>Book of the World's Greatest Teachers</u>, then

Ann Faulkner

McKinley Elementary School

Arlington, Virginia

Will most certainly be one of them.

Her sixth-grade classroom projects and demonstrations got me
started in electronics, a lifelong career.

Contents

Chapter 4

Chapter 5

Chapter 11

Chapter 12

Chapter 13

Chapter 17

Chapter 18

PREFACE

The response to my book *RadioScience Observing, Volume 1* was heartening, to say the least. Readers from all over the world have contacted me, and complimented the book. Some of them made suggestions for material to be included in the second volume, and many of those suggestions are incorporated in this book.

RadioScience Observing is a term I coined for a series of magazine articles in various publications. It surprised me how people reacted to the series, especially since the articles appeared in different magazines and in different countries. But, despite demographic differences between the magazines and cultural differences in different countries, the reaction was the same: very positive.

One of the things that attracts people to RadioScience Observing is that it is true science, but is open to accomplishments by amateur scientists. And some of those accomplishments are considerable. One of the three invited chapters in this book is written by Forrest Mims, winner of the prestigious Rolex Award (which is given to amateur scientists who make a significant contribution). It is also about radio signals, so attracts the same people who are interested in ham radio, shortwave listening, scanner monitoring, and radio propagation.

This volume expands the theme a little bit because of recommendations from the readers of the first volume. You will find chapters on electromagnetic interference (including finding and suppressing same), seismography, ultraviolet radiation, SETI, and geomagnetism.

I would like to thank the authors of three invited chapters: Allan Coleman (Chapter 18, "Long Period Velocity Type Seismometers"), Forrest Mims (Chapter 8, "Monitoring the Ultraviolet Radiation from the Sun"), and Dr. Paul Shuch (Chapter 14, "Setting Up for SETI").

Joseph J. Carr

P.O. Box 1099

Falls Church, VA 22041, USA

E-mail: CARRJJ@AOL.COM

Chapter 1
INTRODUCTION TO
RADIO FREQUENCIES

Chapter 1

Introduction to Radio Frequencies

Radio frequency signals are electromagnetic waves, as opposed to acoustic or mechanical waves, and, for the useful spectrum, generally inhabit the region above 10 kHz or so. However, like many rules of thumb, there are exceptions to the "10 kHz" rule: natural phenomena produce radio waves under 10 kHz ("whistlers" and others), and the U.S. Navy operates a submarine communications network near 60 Hz. Radio frequency (RF) electronics is different because at the higher frequencies, some circuit operation is a little hard to understand or predict. The principal reason is that *stray capacitance* and *stray inductance* afflicts these circuits. Stray capacitance is the capacitance that exists between conductors of the circuit, between conductors or components and ground, or internal to components. Stray inductance is the normal inductance of the conductors that interconnect components, as well as internal component inductances.

These stray parameters are not usually important at low AC frequencies and DC, but as frequency increases they become a much larger proportion of the total. In some older VHF TV tuners and VHF communications receiver front-ends, the stray capacitances were sufficiently large to tune the circuits, so no actual capacitors were needed.

There is also the fact that *skin effect* exists at RF frequencies. The term "skin effect" refers to the fact that AC flows only on the outside portion of the conductor, while DC flows through the entire conductor. As frequency increases, skin effect produces a smaller and smaller zone of conduction...and correspondingly higher value of AC resistance compared with DC resistance.

Another problem with RF circuits is that the signals find it easier to radiate both from the circuit and within the circuit. Thus, coupling effects between elements of the circuit, between the circuit and its environment, and from the environment to the circuit become a whole lot more critical at RF. Interference and other strange effects are found at RF that are missing at DC and negligible in most low-frequency AC circuits.

For these reasons, RF circuits are different from DC and low-frequency AC circuits. When an RF electrical signal radiates, it becomes an *electromagnetic wave*. These waves include not only radio signals, but also infrared (IR), visible light, ultraviolet light (UV), X-rays, gamma rays and others. Before proceeding with our discussion of RF electronic circuits, therefore, we ought to take a look at the electromagnetic spectrum.

The electromagnetic spectrum (*Figure 1-1*) is broken into bands for the sake of convenience and identification, although there is some fuzziness on the exact boundaries of the different bands because different authorities place the band edges at different points. The full spectrum extends from the very lowest AC frequencies and continues well past visible light frequencies into the X-ray and gamma ray region. The *extremely low frequency* (ELF) range includes AC power line frequencies as well as other low frequencies in the 25 to 100 Hertz (Hz) region. The U.S. Navy uses these frequencies for submarine communications. The *very low frequency* (VLF) region extends from just above the ELF region, although most authorities peg it to frequencies from 10 kilohertz (kHz) to 100 kHz; at least one text lists VLF from 10 kHz to only 30 kHz. The *low frequency* (LF) region runs from 100 kHz to 1000 kHz (or 1 MHz), although again some texts put 300 kHz as the upper limit on the LF. The *medium wave* (MW) or *medium frequency* (MF) region runs from 1 MHz to 3 MHz according to some authorities, and 300 kHz to 3 MHz according to others. The AM broadcast band (540 kHz to 1700 kHz)* spans portions of the LF and MW bands. The *high frequency* (HF) region, also called the *shortwave bands* (SW), runs from 3 MHz to 30 MHz. The *very high frequencies* (VHF) start at 30 MHz and run to

*Originally the AM broadcast band in the USA was 540 - 1610 kHz, but that was recently changed to the present 540 - 1700 kHz allocation. Other countries may use slightly different band edges.

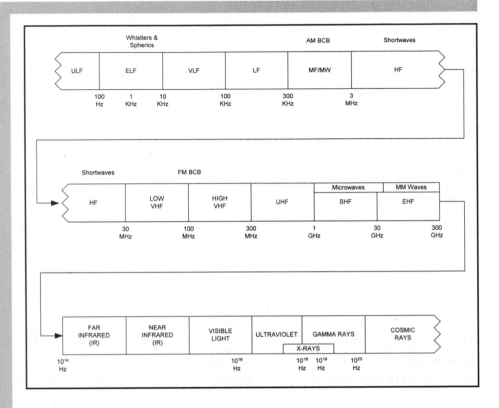

Figure 1-1.
The electromagnetic
spectrum.

300 MHz. This region includes the FM broadcast band, public utilities, some television stations, aviation users, and amateur radio bands. The *ultra high frequencies* (UHF) run from 300 to 900 MHz, and include many of the same services as VHF.

The microwave region officially begins above the UHF region, at 900 or 1000 megahertz (MHz) depending upon source authority. One may well ask how microwaves differ from other electromagnetic waves. Microwaves almost become a separate topic of study in RF because at these frequencies the wavelength approximates the physical size of ordinary electronic components. Thus, components behave differently at microwave frequencies than they do at lower frequencies. At microwave frequencies a half-watt metal film resistor, for example, looks like a complex RLC network with distributed L and C values...and a surprisingly different R value. These tiniest of distributed components have immense significance at microwave frequencies, even though they can be ignored as negligible at lower RF frequencies.

Before examining RF theory, let's first review some background and fundamentals.

Units and Physical Constants

In accordance with standard engineering and scientific practice, all units in this book will be in either the CGS (centimeter-gram-second) or MKS (meter-kilogram-second) systems unless otherwise specified. Because the metric (CGS and MKS) systems depend upon using multiplying prefixes on the basic units, we include a table of common metric prefixes (*Table 1-1*). Other tables are as follows: *Table 1-2* gives the standard physical units; *Table 1-3* shows physical constants of interest in this and other chapters; and *Table 1-4* lists some common conversion factors.

Wavelength and Frequency

For all forms of wave, the velocity, wavelength and frequency (see *Figure 1-2*) are related such that the product of frequency and wavelength is equal to the velocity. For microwaves this relationship can be expressed in the form:

$$c = \lambda F \sqrt{\varepsilon}_0 \qquad\qquad (1\text{-}1)$$

Where:

λ is the wavelength in meters

F is the frequency in hertz (Hz)

ε is the dielectric constant of the propagation medium

c is the velocity of light (300,000,000 m/s)

The dielectric constant (ε) is a property of the medium in which the wave propagates. The value of ε is defined as 1.000 for a perfect vacuum, and very nearly 1.0 for dry air (typically 1.006). In most practical applications the value of ε in dry air is taken to be 1.000. For mediums other

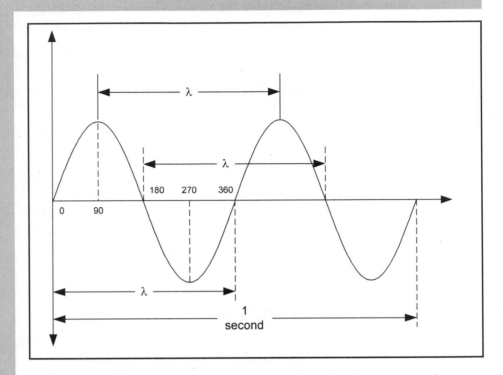

Figure 1-2.
Definition of
frequency and
wavelength.

than air or vacuum, however, the velocity of propagation is slower, and the value of ε relative to a vacuum is higher. Teflon®, for example, can be made with dielectric constant values (ε) from about 2 to 11.

Equation 1-1 is more commonly expressed in the forms of Equations 1-2 and 1-3:

$$\lambda = \frac{c}{F \sqrt{\varepsilon}} \quad 0 \qquad\qquad (1\text{-}2)$$

and,

$$F = \frac{c}{F \sqrt{\varepsilon}} \qquad\qquad (1\text{-}3)$$

(All terms as defined for Equation 1-1).

Microwave Letter Band Designations. During World War II the U.S. military began using microwaves in radar and other applications. For security reasons, alphabetic letter designations were adopted for each band in the microwave region. Because the letter designations became ingrained, they are still used through industry and the defense establishment. Unfortunately, some confusion exists because there are at least three systems currently in use: a) pre-1970 military (*Table 1-5*), b) post-1970 military (*Table 1-6*), and c) an IEEE-Industry standard (*Table 1-7*). Additional confusion is created because the military and defense industry use both pre- and post-1970 designations simultaneously, and industry often uses military rather than IEEE designations. The "old military" designations (*Table 1-5*) persist as a matter of habit.

Skin Effect

There are three reasons why ordinary lumped constant electronic components do not work well at VHF, UHF and microwave frequencies. The first, mentioned earlier in this chapter, is that component size and lead lengths approximate those wavelengths. The second is that distributed values of inductance and capacitance become significant at these frequencies. The third is a phenomenon called *skin effect*.

Skin effect refers to the fact that alternating currents tend to flow on the surface of a conductor. While DC currents flow in the entire cross-section of the conductor, AC flows in a narrow band near the surface. Current density falls off exponentially from the surface of the conductor towards the center (*Figure 1-2*). At the *critical depth* (δ), also called *depth of penetration*, the current density is 1/e, or 1/2.718 = 0.368, of the surface current density. The value of δ is a function of operating frequency, the permeability (μ) of the conductor, and the conductivity (σ). Equation 1-4 gives the relationship.

$$\delta = \sqrt{\frac{1}{2\pi F \sigma \mu}} \qquad (1\text{-}4)$$

Where:

δ is the critical depth

F is the frequency in hertz (Hz)

μ is the permeability in Henrys per meter (H/m)

σ is the conductivity in mhos per meter

Metric Prefixes		
METRIC PREFIX	MULTIPLYING FACTOR	SYMBOL
tera	10^{12}	T
giga	10^9	G
mega	10^6	M
kilo	10^3	K
hecto	10^2	h
deka	10	da
deci	10^{-1}	d
centi	10^{-2}	c
milli	10^{-3}	m
micro	10^{-6}	μ
nano	10^{-9}	n
pico	10^{-12}	p
femto	10^{-15}	f
atto	10^{-18}	a

Table 1-1.
Metric Prefixes.

Units		
QUANTITY	UNIT	SYMBOL
Capacitance	farad	F
Electric charge	coulomb	Q
Conductance	Siemens	S
Conductivity	Siemens/meter	S/m
Current	ampere	A
Energy	joule (watt-sec)	j
Field	volts/meter	E
Flux linkage	weber	(volt-second)
Frequency	hertz	Hz
Inductance	Henry	H
Length	meter	m
Mass	gram	g
Power	watt	W
Resistance	ohm	Ω
Time	second	s
Velocity	meter/second	m/s
Electric Potential	volt	V

Table 1-2.
Units.

Physical Constants		
CONSTANT	VALUE	SYMBOL
Boltzmann's Constant	1.38×10^{-23} J/K	K
Electric Charge (e⁻)	1.6×10^{-19} C	q
Electron (volt)	1.6×10^{-19} J	eV
Electron (mass)	9.12×10^{-31} kg	m
Permeability of Free Space	$4\pi \times 10^{-7}$ H/m	μ_o
Permitivity of Free Space	8.85×10^{-12} F/m	ε_o
Planck's Constant	6.626×10^{-34} J-s	h
Velocity of Electromagnetic Waves	3×10^8 m/s	c
Pi	3.1415927	π

Table 1-3.
Physical Constants.

Conversion Factors

1 inch = 2.54 cm

1 inch = 25.4 mm

1 foot = 0.305 m

1 statute mile = 1.61 km

1 nautical mile = 6,080 feet (6,000 feet)[*]

1 statute mile = 5,280 feet

1 mil = 0.001 inc = 2.54 X 10^{-5} m

1 kg = 2.2 lb

1 neper = 8.686 dB

1 gauss = 10,000 teslas

[*] Navigators use 6,000 feet for ease of calculation. The nautical mile is 1/360 of the earth's circumference at the equator, more or less.

Table 1-4.
Conversion Factors.

BAND DESIGNATION	FREQUENCY RANGE
P	225 - 390 MHz
L	390 - 1550 MHz
S	1550 - 3900 MHz
C	3900 - 6200 MHz
X	6.2 - 10.9 GHz
K	10.9 - 36 GHz
Q	36 - 46 GHz
V	46 - 56 GHz
W	56 - 100 GHz

Table 1-5.
U.S. Military Micro-wave Frequency Bands.

BAND DESIGNATION	FREQUENCY RANGE
A	100 - 250 MHz
B	250 - 500 MHz
C	500 - 1000 MHz
D	1000 - 2000 MHz
E	2000 - 3000 MHz
F	3000 - 4000 MHz
G	4000 - 6000 MHz
H	6000 - 8000 MHz
I	8000 - 10,000 MHz
J	10 - 20 GHz
K	20 - 40 GHz
L	40 - 60 GHz
M	60 - 100 GHz

Table 1-6.
New U.S. Military Microwave Frequency Bands.

BAND DESIGNATION	FREQUENCY RANGE
HF	3 - 30 MHz
VHF	30 - 300 MHz
UHF	300 - 1000 MHz
L	1000 - 2000 MHz
S	2000 - 4000 MHz
C	4000 - 8000 MHz
X	8000 - 12000 MHz
Ku	12 - 18 GHz
K	18 - 27 GHz
Ka	27 - 40 GHz
Millimeter	40 - 300 GHz
Submillimeter	above 300 GHz

Table 1-7. IEEE/Industry Standard Microwave Frequency Bands.

Chapter 2
RADIOSCIENCE RECEIVER BASICS

Chapter 2
RadioScience Receiver Basics

Whatever else you say about radio receivers, one fact remains true: radio signal reception is basically a game of signal-to-noise ratio (SNR). I discussed SNR extensively in *RadioScience Observing Vol. 1*, so will not do so again here. We need to discuss the matter of SNR as it affects receiver system design. As part of this exercise we will make a couple of calculations of noise factor for receivers.

The problem is that a noise signal is seen by any amplifier or receiver following the noise source as a *valid input signal*. Each stage (amp or receiver) in the cascade chain amplifies both the signals *and* noise from previous stages, and also contributes additional noise of its own. Thus, in a cascade amplifier or receiver chain the final stage sees an input signal consisting of the original signal and noise amplified by each successive stage plus the noise contributed by earlier stages. The overall noise factor for a cascade amplifier is calculated from *Friis' noise equation*:

$$F_N \; = \; F_1 \; + \; \frac{F_2 - 1}{G1} \; + \; \frac{F_3 - 1}{G1\,G2} \; + ... + \; \frac{F_N - 1}{G1\,G2 \ldots G_{N-1}} \qquad (2\text{-}1)$$

Where:

   F_n is the overall noise factor of N stages in cascade

   F_1 is the noise factor of stage 1

   F_2 is the noise factor of stage 2

   F_n is the noise factor of the nth stage

   G1 is the gain of stage 1

   G2 is the gain of stage-2

   G_{n-1} is the gain of stage (n-1).

As you can see from Friis' equation, the noise factor of the entire cascade chain is dominated by the noise contribution of the first stage or two. If you want proof, then select some values and "run the numbers."

High-gain, low-noise RF amplifier chains (or receivers) typically use *low noise amplifier* (LNA) circuits for the first stage or two in the cascade chain. Thus, you will find an LNA at the feedpoint of a satellite receiver's dish antenna, and possibly another one at the input of the receiver module itself, but other amplifiers in the chain might be more modest (although their noise contribution cannot be ignored at radio astronomy signal levels).

Receiver Noise Floor

The *noise floor* of the receiver is a statement of the amount of noise produced by the receiver's internal circuitry, and directly affects the *sensitivity* of the receiver. The noise floor is typically expressed in dBm. The noise floor specification is evaluated as follows: the more negative the better. The best receivers have noise floor numbers of less than -130 dBm, while some very good commercial receivers of moderate cost offer numbers of -115 dBm to -130 dBm (note: some radio astronomy receiver systems have values around -200 dBm).

The noise floor is directly dependent on the bandwidth used to make the measurement. Receiver advertisements usually specify the bandwidth, but note whether or not the bandwidth that produced the very good performance numbers is also the bandwidth that you'll need for the mode of transmission you want to receive. If, for example, you are interested only in weak 6 kHz-wide AM signals, and the noise floor is specified for a 250 Hz CW filter, then the noise floor might be too high for your use.

Receiving System Example

Figure 2-1 shows a receiving system that is common in the VHF through microwave regions of the spectrum. An antenna is used to obtain the signal, and a *low-noise amplifier* (LNA), A1 in *Figure 2-1*, is provided to boost the antenna signal. It is common practice to place the LNA at the antenna terminals (e.g., at feed horn port) so that is does not have to overcome the loss of the transmission line. The receiver may or may not have an RF amplifier, but in this model an RF amplifier (A2) is used. The mixer is then used to convert the RF signal to the IF used by the receiver.

Loss of the coaxial cable transmission line can be a significant cause of noise in the system. The cable loss is usually expressed in decibels (dB), and is taken from the manufacturer's data sheets if no actual measurements are available. Typically, the manufacturer will provide a chart that relates loss in *decibels per meter* (dB/m) or *decibels per hundred feet* (dB/100-ft) to frequency.

Find the loss factor appropriate to the desired frequency, and correct for the actual length of the line. The noise temperature of the transmission line is:

$$T_{e(line)} = T_L(L - 1) \tag{2-2}$$

Where:

$T_{e(line)}$ is the noise temperature of the line

L is the loss of the line expressed in linear terms (as a ratio)

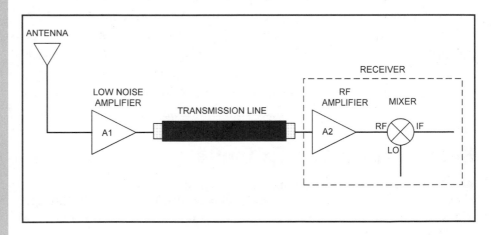

Figure 2-1.
Receiver front-end
block diagram.

Table 2-1 shows the results of making the noise calculations on a receiving system such as *Figure 2-1* when the following specifications are used:

Stage	Gain (dB)	Noise Figure (dB)
Preamp	15	2.2
Trans. Line	-2	2.0
RF Amp	10	3.0
Mixer	-6	4.5

BANDWIDTH (Hz)	NOISE ($\times 10^{-8}$ Volts)
1,000	2.83
1,500	3.46
2,000	4.00
2,500	4.47
3,000	4.90
3,500	5.29
4,000	5.66
4,500	6.00
5,000	6.33
5,500	6.63
6,000	6.93
6,500	7.21
7,000	7.49
7,500	7.75
8,000	8.00
8,500	8.25
9,000	8.49
9,500	8.72
10,000	8.95

Table 2-1.
Noise and
bandwidth

The overall gain for this portion of the receiver is the sum of the gains, or 17 dB. The results of the Friis equation shows an overall noise figure of 2.398. If you program a spreadsheet with the noise equations so that you can vary the noise figure parameters, it becomes apparent that the first stage dominates. Let's do a little *ceterus paribus** exercise in which one noise figure is changed by 1 dB. If the preamplifier noise figure is increased to 3.2 dB, then the overall noise figure rises to 3.36 dB. Increasing the transmission line noise figure to 3 dB only raises the noise figure to 2.47 dB. Increasing the RF amplifier noise figure to 4 dB increases the overall noise figure to 2.46 dB. Finally, increasing the mixer noise figure to 2.41 dB, let's tabulate the results:

*All else remaining unchanged.

For a 1-dB increase in noise figure, the overall noise figure changes to:

Stage	New N.F. (dB)	Change
Low-noise amplifier	3.20	+0.8 dB
Transmission line	2.47	+0.072 dB
RF amplifier	2.46	+0.062 dB
Mixer	2.41	+0.012 dB

Note that the increase in overall noise figure is greatest for the first stage in the chain, and that the change for each succeeding stage is less than for the stage before. The lesson here is to put as much effort as possible into the first stage in order to reduce the noise figure overall.

The bottom line is simple: Use low-noise stages throughout your system, but especially in the very first stage. The result of adding a noisy preamplifier to the front-end of the system will be to reduce the overall noise figure. This occurs because the *overall system noise figure is absolutely dominated by the noise figure of the first stage*.

Noise Equations

These equations allow you to calculate the thermal noise, noise figure, noise factor and noise temperature of systems. They are discussed in *RadioScience Observing, Volume 1*.

Thermal Noise:

$$V_N \;=\; \sqrt{4\,K\,T\,B\,R} \tag{2-3}$$

Where:

> V_N is the noise potential in volts (V)

> K is Boltzmann's constant (1.38 x 10^{-23} J/°K)

> T is the temperature in degrees Kelvin (°K), normally set to an average "room temperature" of 290 °K by convention.

> R is the resistance in ohms (Ω)

> B is the bandwidth in hertz (Hz)

Thermal Noise as a function of bandwidth:

$$V_N \;=\; \sqrt{4\,K\,T\,R} \quad V/\sqrt{Hz} \tag{2-4}$$

Noise Factor:

$$F_N \;=\; \left[\frac{P_{NO}}{P_{NI}} \right]_{T\,=\,290\ °K} \tag{2-5}$$

Noise Figure:

$$N.F. \;=\; 10\,LOG\,F_N \tag{2-6}$$

Noise Temperature:

The noise temperature is related to the noise factor by:

$$T_e = (F_N - 1)T_o \tag{2-7}$$

and to noise figure by

$$T_e = 290 \left[10^{(N.F./10)} - 1 \right] \tag{2-8}$$

Chapter 3
SOME USEFUL
RECEIVER CIRCUITS

Chapter 3
Some Useful Receiver Circuits

In this chapter we will look at several useful receiver circuits for those who wish to "roll their own" RadioScience Observing receivers. *Figure 3-1* shows the basic block diagram for a standard radio receiver. This particular design is called a *superheterodyne*, and even though it dates from the 1920s is still the design of choice. The "heterodyne" comes from the fact that this type of receiver converts the incoming RF frequency to some other fixed frequency called the *intermediate frequency* (IF) by mixing it with a *local oscillator* signal (LO).

The first stage after the antenna in *Figure 3-1* is the preamplifier. This stage may or may not be tuned, depending on the design. For high-quality radio receivers this amplifier must be a low-noise amplifier (LNA), and for RadioScience Observing receivers the use of an LNA is essential.

The *mixer* stage is the heart of the superheterodyne receiver. It receives the LO and RF signals, and combines them nonlinearly to produce an output spectrum that consists of *sum* (LO+RF) and *difference* (LO-RF) frequency signals. Depending on the type of mixer used, the LO and/or RF may or may not be present in the mixer output. In double-balanced mixers, the kind used for most RadioScience Observing receivers, neither the RF nor the LO will appear in the mixer output.

The IF amplifier stage provides the largest amount of gain and most of the selectivity of the receiver. This stage essentially supplies the amplification of the overall receiver.

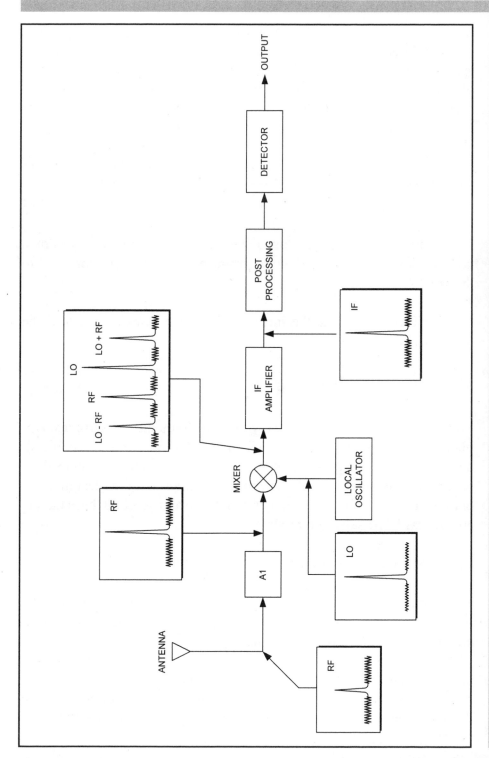

Figure 3-1.
Block diagram of a
superheterodyne
receiver.

In some receivers, there is some form of post-processing taking place. This might be additional filtering, or it might be sorting out signals by spectrum analysis. In other cases, in order to increase the dynamic range of the receiver, or to convert the output to dBm power units, the post-processor will consist of a logarithmic amplifier. An example is provided later in this chapter.

Finally, there is a detector or demodulator stage that is used to recover any information on the signal. For example, a simple envelope detector is used for AM signals, a discriminator (of several types) is used for FM signals, and a product detector for SSB signals. In many RadioScience Observing receivers an envelope detector is used, but is followed by a time-averaging integrator stage to produce a DC level proportional to the input signal level.

Now let's take a look at the mixer circuits...the heart and soul of the receiver system.

RF Receiver Mixers

Mixer circuits are used extensively in radio-frequency electronics. Applications include frequency translators (including in radio receivers), demodulators, limiters, attenuators, phase detectors and frequency doublers. There are a number of different approaches to mixer design. Each of these approaches has advantages and disadvantages, and these factors are critical to the selection process.

Linear vs. Nonlinear Mixers

The word "mixer" is used to denote both linear and nonlinear circuits. And this situation is unfortunate because only the nonlinear is appropriate for the RF mixer applications listed above.

So what's the difference? The basic linear mixer is actually a *summer* circuit, as shown in *Figure 3-2A* (the schematic symbol is in *Figure 3-2B*). Some sort of combiner is needed. In the case shown, the com-

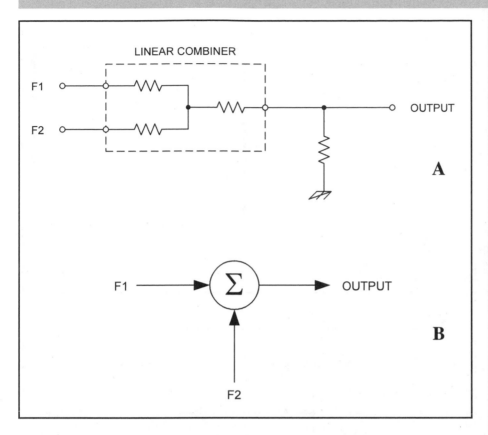

LINEAR COMBINER

F1

F2

OUTPUT

A

F1 ────▶ (Σ) ────▶ OUTPUT

B

F2

Figure 3-2.
A) Linear "mixer"
circuit; B) symbol.

biner is a resistor network. There is no interaction between the two input signals, F1 and F2. They will share the same pathway at the output, but otherwise do not affect each other. This is the action one expects of microphone and other audio mixers. If you examine the output of the summer on a spectrum analyzer (*Figure 3-3*), you will see the spikes representing the two frequencies, and nothing else other than noise.

The nonlinear mixer is shown in *Figure 3-4A*, and the circuit symbol in *Figure 3-4B*. While the linear mixer is a summer, the nonlinear mixer is a *multiplier*. In this particular case, the nonlinear element is a simple diode, such as a 1N4148 or similar devices. Mixing action occurs when the nonlinear device, such as diode D1, exhibits impedance changes over cyclic excursions of the input signals. In order to achieve switching action, one signal must be considerably higher than the other. It is commonly assumed that a 20 dB or more difference is necessary.

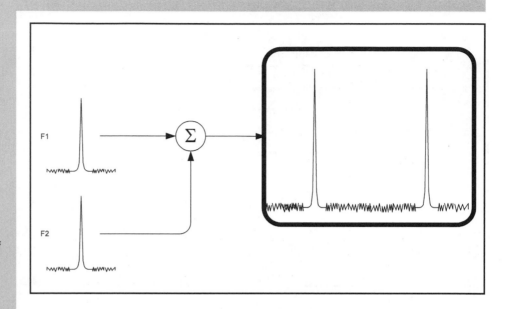

Figure 3-3.
Spectrum display of
the output of a
linear mixer.

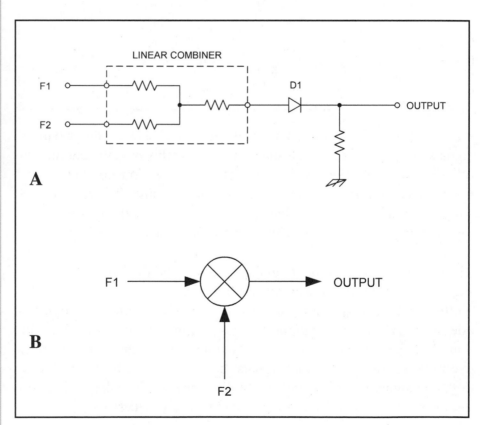

Figure 3-4.
A) Nonlinear mixer
circuit;
B) Symbol.

Whenever a nonlinear element is added to the signal path a number of new frequencies will be generated. If only one frequency is present, then we would still expect to see its harmonics; for example, F1 and nF1 where n is an integer. But when two or more frequencies are present, a number of other products are also present. The output frequency spectrum from a nonlinear mixer is:

$$\pm F_o = mF1 \pm nF2 \qquad\qquad (3\text{-}1)$$

Where:

F$_O$ is the output frequency for a specific (m,n) pair

F1 and F2 are the applied frequencies

m and n are integers or zero (0, 1, 2, 3...).

There will be a unique set of frequencies generated for each (m, n) ordered pair. These new frequencies are called *mixer products* or *intermodulation products*. *Figure 3-5* shows how the output would look on a spectrum analyzer. The original signals (F1 and F2) are present, along with an array of mixer products arrayed at frequencies away from F1 and F2.

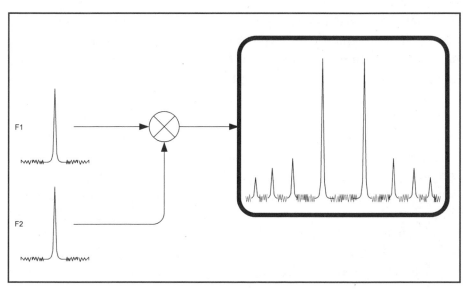

Figure 3-5.
Spectrum display of the output of a nonlinear mixer.

The implication of *Equation 3-1* is that there will be a large number of (m, n) frequency products in the output spectrum. Not all of them will be useful for any specific purpose, and may well cause adverse effects.

So why do we need mixers? There are other ways to generate various frequencies, so why a frequency translator such as a heterodyne mixer? The principal answer is that the mixer will translate the frequency, and in the process transfer the modulation of the original signal. So, when an AM signal is received, and then translated to a different frequency in the receiver, the modulation characteristics of the AM signal convey to the new frequency essentially undistorted (those who know that there are no "distortionless" circuits please refrain from snickering). Perhaps the most common use for mixers, in this regard, is in radio receivers.

The Receiver Mixer

The vast majority of radio receivers made since the late 1920s have been *superheterodynes*. The process of heterodyning is the translation of one frequency to another by the use of a mixer and local oscillator. *Figure 3-6A* shows this application of mixer in block diagram form (this is an expansion of *Figure 3-1*).

The antenna picks up a radio signal of frequency F_{RF}, and mixes it with a local oscillator signal F_{LO}. This produces a number of new frequencies in the spectrum defined by *Equation 3-1*, but those of principal interest are the cases where (m, n) = (1, 1); i.e., the *sum* and *difference* frequencies $F_{RF} + F_{LO}$ and $F_{RF} - F_{LO}$.

One of these second-order products will be selected by an IF filter, and the other is rejected. Why would receiver designers use this approach? The principal reason is that it is very much easier to design the receiver using this approach. It is much easier to provide the gain and selectivity filtering needed to make the receiver work properly at a single frequency. This frequency, regardless of whether the sum or difference product is used, is called the *intermediate frequency* (IF), or F_{IF}. The high-gain stages, and the bandpass filtering, are all provided in the IF stages.

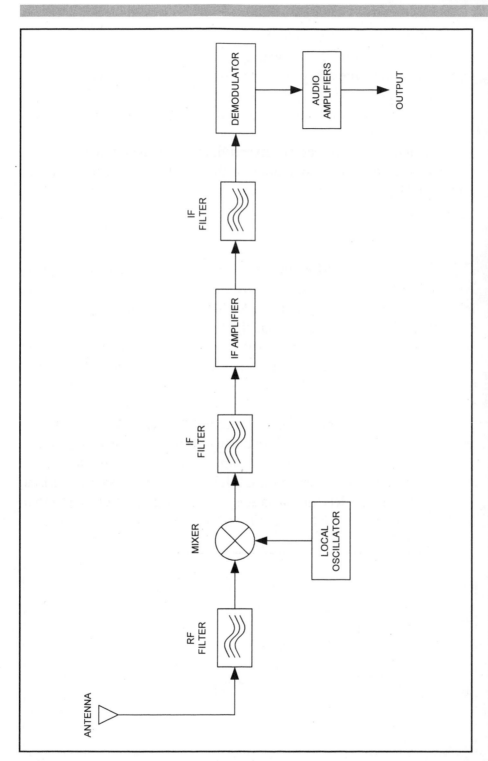

Figure 3-6A.
Block diagram of a
superheterodyne
receiver.

Terminology. In the remainder of this chapter F1 and F2 will be expressed much of the time as F_{RF} and F_{LO} in view of the receiver being the most common use for mixer devices.

At one time, it was universally the practice to select the difference frequency, but today the sum frequency is often selected. It is quite common to find high-frequency (HF) shortwave receivers with a dual-conversion scheme in which F_{RF} is first up-converted to the sum frequency, and then a new mixer down-converts it to a lower second IF frequency.

The sum or difference second-order products are selected for the IF, but the other frequencies don't simply evaporate. They can cause serious problems. But more of that later.

Simple Diode Mixer

Figure 3-6B shows a block diagram circuit for a simple form of mixer. Although not terribly practical in most cases, the circuit has been popular in a number of receivers in the high UHF and microwave regions since World War II. The two input signals are the RF and LO. The LO signal is at a very much higher level than the RF signal, and is used to switch the diode in and out of conduction, providing the nonlinearity that mixer action requires.

There are three filters shown in this circuit. The RF and LO filters are used for limiting the frequencies that can be applied to the mixer. In the case of the RF port it is other radio signals on the band that are being suppressed, while in the case of the LO it is LO noise and harmonics that are suppressed. The RF filter also serves to reduce any LO energy that may be transmitted back towards the RF input.

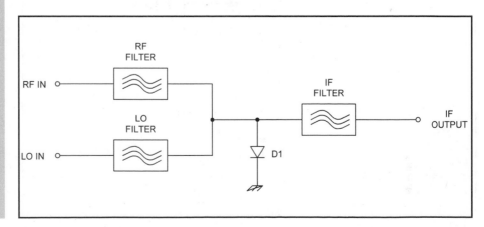

Figure 3-6B.
Basic single-ended
unbalanced mixer
circuit.

There are a number of possibly apocryphal legends from World War II of receiver LO radiation back through the antenna circuit being responsible for an enemy detecting the location of the receiver, so this effect is rather important. One such legend is from British airborne radar history. According to one source of doubtful authority, German submarines sailing on the surface learned to listen for *Beaufighter* centimetric radars using a receiver that was poorly suppressed. The aircrews then learned that they could locate the submarine with just the radar's receiver tuned to listen for the submarine receiver's LO [anyone with first-hand knowledge of this matter, please contact the author].

The Question of "Balance"

One of the ways of classifying mixers is whether or not they are *unbalanced*, *single-balanced* or *double-balanced*. Although there are interesting aspects of each of these categories, we are presently interested in how they affect the output spectrum.

Unbalanced Mixers: Both F_{RF} and F_{LO} appear in the output spectrum, and there may be poor LO-RF and RF-LO port isolation. Their principal attraction is low cost.

Single-Balanced Mixers: Either F_{RF} or F_{LO} is suppressed in the output spectrum, but not both (i.e., if F_{RF} is suppressed, F_{LO} will be present, and vice versa). The single-balanced mixer will also suppress even-order LO harmonics ($2F_{LO}$, $4F_{LO}$, $6F_{LO}$, etc.). High LO-RF isolation is provided, but LO-IF isolation must be provided by external filtering.

Double-Balanced Mixers: Both F_{RF} and F_{LO} are suppressed in the output. The single-balanced mixer will also suppress even-order LO and RF harmonics ($2F_{LO}$, $2F_{RF}$, $4F_{LO}$, $4F_{RF}$, $6F_{LO}$, $6F_{RF}$, etc.). High port-to-port isolation is provided.

Spurious Responses

The IF section of a receiver will use one of the second-order products in order to convert F_{RF} to F_{IF}. Ideally, the receiver would only respond to the single RF frequency that meets the need. Unfortunately, reality some-times rudely intervenes, and certain spurious responses might be noted.

A *spurious response* in a superheterodyne receiver is any response to any frequency other than the desired F_{RF}, and which is strong enough to be heard in the receiver input. Most of these "spurs" are actually mixer responses, although overloading the RF amplifier can cause some re-sponses as well. The mixer responses may or may not be affected by pre-mixer filtering of the RF signal. Candidate spur frequencies include any that satisfy *Equation 3-2*:

$$F_{Spur} = \frac{n F_{LO} \pm F_{IF}}{m} \qquad\qquad (3\text{-}2)$$

Image. The image response of a mixer is due to the fact that two fre-quencies satisfy the criteria for F_{IF}. *Figure 3-6* shows how the image response works. The frequency that satisfies the image criteria depends on whether the LO is *high-side injected* ($F_{LO} > F_{RF}$) or *low-side injected* ($F_{LO} < F_{RF}$). In the high-side injection case [(m, n) = (1, -1)] shown in *Figure 3-7*, the image appears at $F_{RF} + 2F_{IF}$. If low-side injection [(m, n) = (-1, 1)] is used, then the image is at $F_{RF} - 2F_{IF}$. The image always appears on the *opposite side of the LO from the RF*, so will be $F_{LO} + F_{IF}$ for high-side injection and $F_{LO} - F_{IF}$ for low-side injection.

Let's consider an actual example based on an AM *broadcast band* (BCB) receiver. The IF is 455 kHz, and the receiver is tuned to $F_{RF} = 1{,}000$ kHz. The usual procedure on AM BCB receivers is high-side injection, so $F_{LO} = F_{RF} + F_{IF} = 1{,}000$ kHz + 455 kHz = 1,455 kHz. The image fre-quency appears at $F_{RF} + 2F_{IF} = 1{,}000$ kHz + (2 x 455 kHz) = 1,910 kHz. Any signal on or near 1,910 kHz that makes it to the mixer RF input port will be converted to 455 kHz along with the desired signals.

The problem is complicated by the fact that it is not just actual signals present at the image frequency, but noise as well. The noise applied to the mixer input is essentially doubled if the receiver has any significant

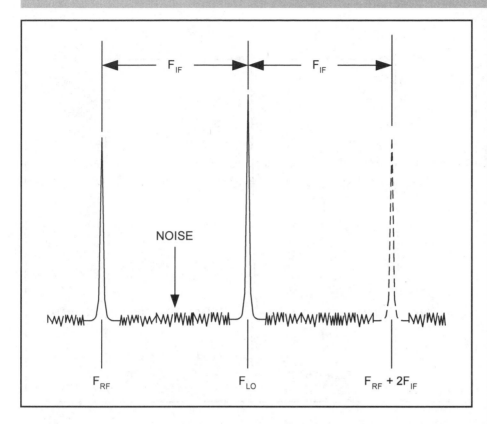

F$_{IF}$ F$_{IF}$

NOISE

F$_{RF}$ F$_{LO}$ F$_{RF}$ + 2F$_{IF}$

Figure 3-7.
Image frequency
for high-side
injection mixer.

response at the image frequency. Premixer filtering is needed to reduce the noise. Receiver designers also specify high IF frequencies in order to move the image out of the passband of the RF prefilter.

Half-IF. Another set of images occurs when (m, n) is (2, -2) for low-side or (-2, 2) for high-side. This image is called the *half-IF image*, and is illustrated in *Figure 3-8*. An interesting aspect of the half-IF image is that it is created by internally generated harmonics of both F$_{RF}$ and F$_{LO}$. For our AM BCB receiver where F$_{RF}$ = 1,000 kHz, F$_{LO}$ = 1,4500 kHz and F$_{IF}$ = 455 kHz, the half-IF frequency is $1,000 + (455/2) = 1,227.5$ kHz.

IF Feedthrough. If a signal from outside passes through the mixer to the IF amplifier, and happens to be on a frequency equal to F$_{IF}$, then it will be accepted as a valid input signal by the IF amplifier. The mixer RF-IF port isolation is critical in this respect.

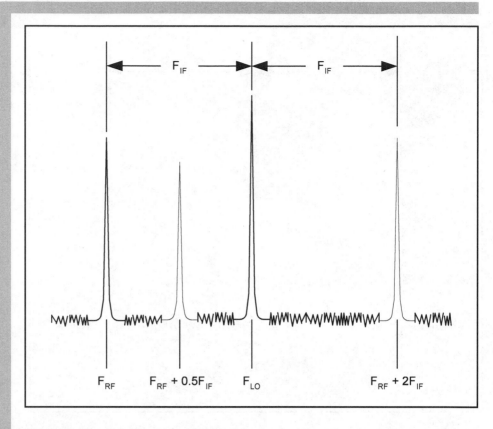

Figure 3-8.
Half-IF response.

High-Order Spurs. Thus far we have considered only the case where a single RF frequency is applied to the mixer. But what happens when two RF frequencies (F_{RF1} and F_{RF2}) are applied simultaneously. This is the actual situation in most practical receivers. There are a large number of higher-order responses (i.e., where m and n are both greater than 1) defined by $mF_{RF1} \pm nF_{RF2}$.

The worst case is usually the ($2F_{RF1}$ - F_{RF2}) and ($2F_{RF2}$ - F_{RF1}) third-order products because they fall close to F_{RF1} and F_{RF2} and may be within the device passband. Although any of the spurs may prove difficult to handle in some extreme cases, the principal problems occur with the third-order difference products of two RF signals applied to the RF port of the mixer ($2F_{RF1}$ - F_{RF2} and $2F_{RF2}$ - F_{RF1}). *Figure 3-9* illustrates this effect for our AM BCB receiver. Suppose two signals appear at the mixer input: F_{RF1} = 1,000 kHz and F_{RF2} = 1,020 kHz (This combination is highly likely in the crowded AM BCB!). The third-order products of these two sig-

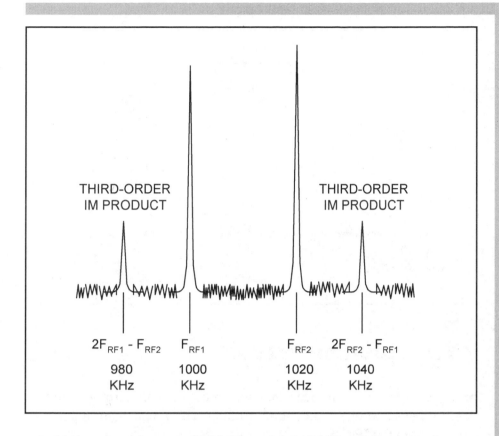

THIRD-ORDER
IM PRODUCT

THIRD-ORDER
IM PRODUCT

$2F_{RF1} - F_{RF2}$ F_{RF1} F_{RF2} $2F_{RF2} - F_{RF1}$

980
KHz

1000
KHz

1020
KHz

1040
KHz

Figure 3-9.
Third-order
products close to
test signals.

nals hitting the mixer are 980 kHz and 1,040 kHz, and appear close to F_{RF1} and F_{RF2}. If the premixer filter selectivity is not sufficiently narrow to suppress the unwanted RF frequency, then the receiver may respond to the third-order products as well as the desired signal.

LO Harmonic Spurs. If the harmonics of the local oscillator are strong enough to drive mixer action, then signals clustered at $\pm F_{IF}$ from each significant harmonic will also cause mixing. *Figure 3-10* shows this effect. The passband of the pre-mixer filter is shown as dotted line curves at $F_{LO} \pm F_{IF}$, $2F_{LO} \pm F_{IF}$ and $3F_{LO} \pm F_{IF}$.

LO Noise Spurs. All oscillators have noise close to the LO frequency. The noise may be due to power supply noise modulating the LO, or it may be random phase noise about the LO. In either case, the noise close to the LO, and within the limits imposed by the IF filter, will be passed through the mixer to the IF amplifier.

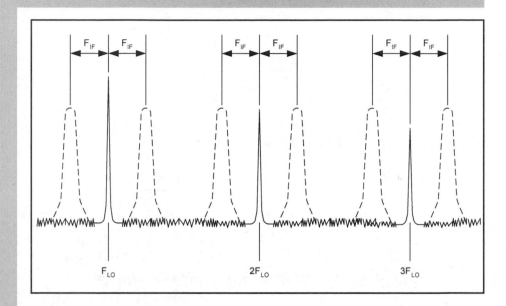

Figure 3-10. Noise balance deteriorates if LO harmonics are present.

Mixer Distortion Products

Because mixers are nonlinear, they will produce both harmonic distortion products and intermodulation products (IPs). Our main interest at this point are the IPs.

Intermodulation Products (IP)

The spurious IP signals generated when two signals, F1 and F2, are mixed nonlinearly are shown graphically in *Figure 3-11* (assuming input frequencies of 1 MHz and 2 MHz). Given input signal frequencies of F1 and F2, the main IPs are:

$$\text{Second-order: } F1 \pm F2$$

$$\text{Third-order: } \quad 2F1 \pm F2$$

$$2F2 \pm F1$$

$$\text{Fifth-order: } \quad 3F1 \pm 2F2$$

$$3F2 \pm 2F1$$

The second-order and third-order products are those normally specified in a receiver mixer design because they tend to be the strongest. In general, even-order IMD products (2, 4, etc.) tend to be less of a problem because they can often be ameliorated by using external filtering ahead of the receiver mixer (or tuned RF amplifier if one is used). Prefiltering tends to reduce the amplitude of out-of-channel interfering signals, reducing the second-order products within the channel. Third-order IMD products are more important because they tend to reflect on the receiver's dynamic range, and its ability to handle strong signals. The third-order products are usually not easily influenced by external filtering, so must be handled by proper mixer selection and/or design.

When an amplifier or mixer is overdriven, the second-order content of the output signal increases as the square of the input signal level, while the third-order responses increase as the cube of the input signal level. When expressed in decibels (dB) the third-order transfer function has a slope three times that of the first-order transfer function (*Figure 3-12*).

Consider the case where two HF signals (F1 = 10 MHz; F2 = 15 MHz) are mixed together. The second-

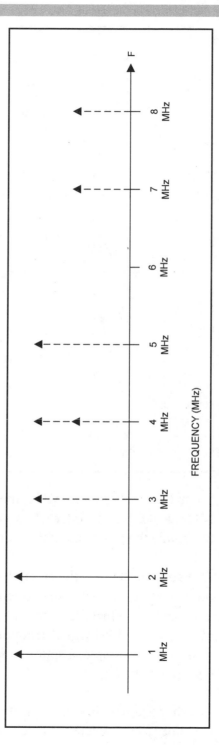

Figure 3-11.
Mixer products
when F1 = 1 MHz
and F2 = 2 MHz.

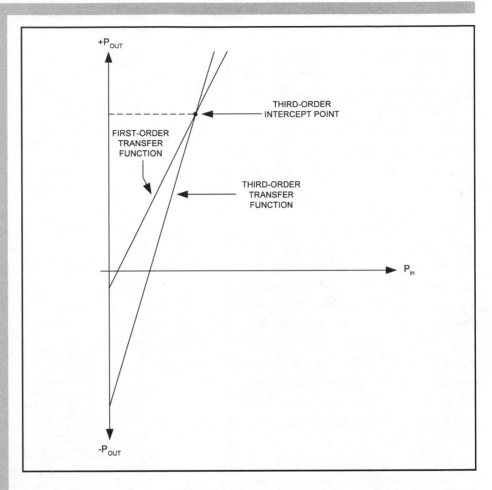

Figure 3-12.
Distortion products.

order IPs are 5 and 25 MHz; the third-order IPs are 5, 20, 35 and 40 MHz; and the fifth-order IPs are 0, 25, 60 and 65 MHz. If any of these are inside the passband of the receiver, then they can cause problems.

One such problem is the emergence of "phantom" signals at the IP frequencies. This effect is seen often when two strong signals (F1 and F2) exist and can affect the front-end of the receiver, and one of the IPs falls close to a desired signal frequency, F_d. If the receiver was tuned to 5 MHz, for example, a spurious signal would be found from the F1-F2 pair given above.

Another example is seen from strong in-band, adjacent channel signals. Consider a case where the receiver is tuned to a station at 9610 kHz, and there are also very strong signals at 9600 kHz and 9605 kHz. The near (in-band) IP products are:

Third-order: 9595 kHz (ΔF = 15 kHz)

9610 kHz (ΔF = 0 kHz)   [ON CHANNEL!]

Fifth-order: 9590 kHz (ΔF = 20 kHz)

9615 kHz (ΔF = 5 kHz)

Note that one third-order product is on the same frequency as the desired signal, and could easily cause interference if the amplitude is sufficiently high. Other third- and fifth-order products may be within the range where interference could occur, especially on receivers with wide bandwidths.

The IP orders are theoretically infinite because there are no bounds on either m or n. However, in practical terms, because each successively higher-order IP is reduced in amplitude compared with its next lower-order mate, only the second-order, third-order and fifth-order products usually assume any importance. Indeed, only the third-order is normally used in receiver specifications sheets.

Third-Order Intercept Point

It can be claimed that the third-order intercept point (TOIP) is the single most important specification of a mixer's dynamic performance, because it predicts the performance in regards to intermodulation, cross modulation and blocking desensitization.

When a mixer is used in a receiver, the third-order (and higher) intermodulation products (IP) are normally very weak, and don't exceed the receiver noise floor when the mixer and any preamplifiers are operating in the linear region. But as input signal levels increase, forcing the front-end of the receiver toward the saturated nonlinear region, the IPs emerge from the noise (*Figure 3-13*) and begin to cause problems. When this happens, new spurious signals appear on the band and self-generated interference arises.

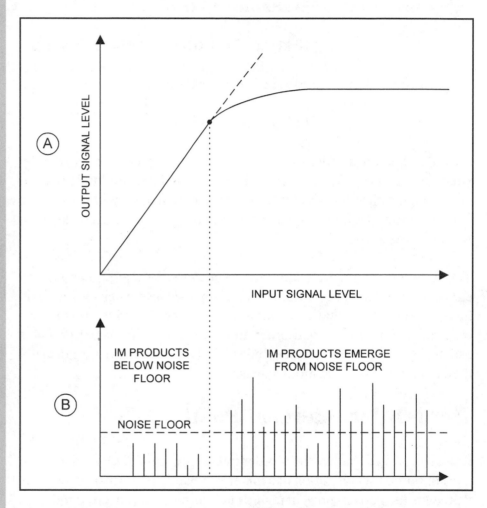

Figure 3-13.
Intermod products
rise out of the noise
when critical input
level is exceeded.

Figure 3-14 shows a plot of the output signal vs. fundamental input signal. Note the output compression effect that occurs as the system begins to saturate. The dotted gain line continuing above the saturation region is the theoretical output that would be produced if the gain did not clip.

It is the nature of third-order products in the output signal to emerge from the noise at a certain input level, and increase as the cube of the input level. Thus, the slope of the third-order line increases 3 dB for every 1-dB increase in the response to the fundamental signal. Although the output response of the third-order line saturates similarly to that of the fundamental signal, the gain line can be continued to a point where it intersects the gain line of the fundamental signal. This point is the *third-order intercept point* (TOIP).

Notice in *Figure 3-14* that the gain (P_O/P_{IN}) begins to decrease in the vicinity of the TOIP. The measure of this tendency to saturation is called the *-1 dB compression point*; i.e., the point where the gain slope decreases by 1 dB.

Interestingly enough, one tactic that can help reduce IP levels back down under the noise is the use of an attenuator ahead of the mixer. Even a few dB of input attenuation is often sufficient to drop the IPs back into the noise, while afflicting the desired signals only a small amount. Many modern receivers provide a switchable attenuator ahead of the mixer.

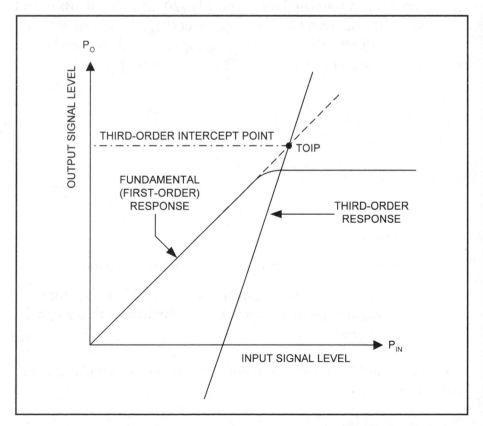

Figure 3-14.
Third-order intercept
point (TOIP).

This practice must be evaluated closely, however, if low-level signals are to be handled. The usual resistive attenuator pad will increase the thermal noise level appearing at the input of the mixer by an amount proportional to its looking-back resistance.

The IP performance of the mixer selected for a receiver design can profoundly affect the performance of the receiver. For example, the second-order intercept point affects the half-IF spur rejection, while the third-order intercept point will affect the intermodulation distortion (IMD) performance.

Calculating Intercept Points

Calculating the nth order intercept point can be done using a two-tone test scheme. A test system is created in which two equal-amplitude signals (F_A and F_B) are applied simultaneously to the mixer RF input. These signals are set to a standard level (typically -20 dBm to -10 dBm), and the power of the nth intermodulation product (P_{IMN}) is measured (using a spectrum analyzer or, if the spectrum analyzer is tied up elsewhere, a receiver with a calibrated S-meter). The nth intercept point is:

$$IP_N = \frac{NP_A - P_{IMN}}{N-1}$$

(3-3)

Where:

IP_N is the intermod product of order N

N is the order of the intermod product

P_A is the input power level (in dBm) of one of the input signals

P_{IMN} is the power level (in dBm) of the nth IM product (often specified in terms of the receiver's minimum discernible signal specification)

Once the P_A and P_{IM} points are found, any IP can be calculated using *Equation 3-3*.

Mixer Losses

Depending on its design, a mixer may show either loss or gain. The principal loss is *conversion* loss, which is made up of three elements: *mismatch loss*, *parasitic loss* and *junction loss* (assuming a diode mixer). The conversion loss is simply a measure of the ratio of the RF input signal level and the signal level appearing at the IF output (P_{IF}/P_{RF}). In some cases, it may be a gain, but for many—perhaps most—mixers there is a loss. Conversion loss (L_C) is:

$$L_C = L_M + L_P + L_J \qquad\qquad (3\text{-}4)$$

Where:

L_C is conversion loss

L_M is the mismatch loss

L_P is the parasitic loss

L_J is the junction loss

Mismatch loss is a function of the impedance match at the RF and IF ports. If the mixer port impedance (Z_P) and the source impedance (Z_S) are not matched, then a VSWR will result that is equal to the ratio of the higher impedance to the lower impedance (VSWR = Z_P/Z_S or VSWR = Z_S/Z_P, depending on which ratio is ≥ 1). The mismatch loss is the sum of RF and IF port mismatch losses. Or expressed in terms of VSWR:

$$L_M = 10 \times \left[LOG_{10}\left[\frac{(VSWR_{RF} + 1)^2}{4\,VSWR_{RF}} \right] + LOG_{10}\left[\frac{(VSWR_{IF} + 1)^2}{4\,VSWR_{IF}} \right] \right]$$

$$(3\text{-}5)$$

Parasitic loss is due to action of the diode's parasitic elements; i.e., series resistance (R_S) and junction capacitance (C_J). Junction loss is a function of the diode's *I-vs.-V* curve. The latter two elements are controlled by careful selection of the diode used for the mixer.

Noise Figure

Radio reception is largely an issue of *signal-to-noise ratio* (SNR). In order to recover and demodulate weak signals, the noise figure (NF) of the receiver is an essential characteristic. The mixer can be a large contributor to the overall noise performance of the receiver. Indeed, the noise performance of the receiver is seemingly affected far out of proportion to the actual noise performance of the mixer. But a study of signals and noise will show (through Friis' equation) that the noise performance of a receiver or cascade chain of amplifiers is dominated by the first two stages, with the first stage being so much more important than the second stage.

Because of the importance of mixer noise performance, a low-noise mixer must be designed or procured. In general, the noise figure of the receiver-equipped diode mixer first stage (as is common in microwave receivers) is:

$$\text{N.F.} = L_C + \text{IF}_{N.F.} \tag{3-6}$$

Where:

$\quad$ N.F. is the overall noise figure

$\quad$ L_C is the conversion loss

$\quad$ $\text{IF}_{N.F.}$ is the noise figure of the first IF amplifier stage.

To obtain the best overall performance from the perspective of the mixer, the following should be observed:

$\quad$ 1. Select a mixer diode with a low noise figure (this will address the junction and parasitic losses).

2. Ensure the impedance match of all mixer ports.

3. Adjust the LO power level for minimum conversion loss (LO power is typically higher than maximum RF power level).

Noise Balance

There is noise associated with the LO signal, and that noise can be transferred to the IF in the mixing process. The tendency of the mixer to transfer AM noise to the IF is called its *noise balance*. In some cases, this transferred noise results in loss that is more profound than the simple conversion loss, so should be evaluated when selecting a mixer.

The total noise picture (*Figure 3-15*) includes not simply the AM noise sidebands around the LO frequency, but also the noise sidebands around the LO harmonics. The latter can be eliminated by imposing a filter between the LO output and the mixer's LO input. The noise sidebands around the LO itself, however, are not easily suppressed by filtering because they are close in frequency to F_{LO}. The use of a balanced mixer, however, can suppress all of the LO signal in the output, and that includes the noise sidebands. In the usual way noise balance is specified, the higher the number (in dB) the more suppression of LO AM noise.

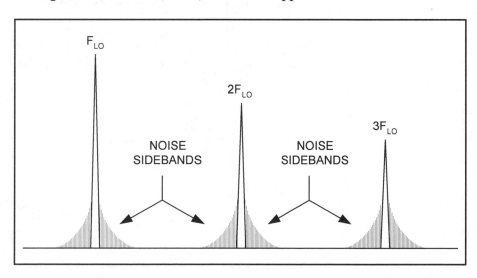

Figure 3-15.
Noise sidebands surrounding LO and its harmonics can deteriorate sensitivity.

Single-Ended Active Mixer Circuits

Thus far the only mixer circuit that has been explicitly discussed is the diode mixer. The diode is a general category called a *switching mixer* because the LO switches the diode in and out of conduction. Now let's turn our attention to active single-ended unbalanced mixers.

Figure 3-16 shows the circuit of a simple single-ended unbalanced mixer based on a junction field-effect transistor (JFET) such as the MPF-102

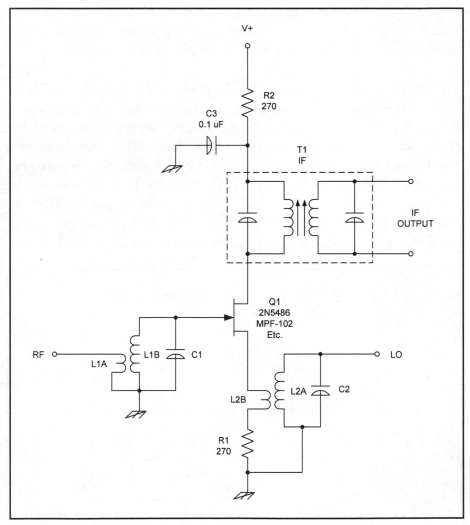

Figure 3-16.
Single-ended JFET
mixer.

or 2N5486. The RF signal is applied to the gate, while the LO signal is applied to the source. If the LO signal has sufficient amplitude to cause nonlinear action, then it will permit the JFET to perform as a mixer.

Note that both the RF and LO ports are fitted with bandpass filters to limit the frequencies that can be applied to the mixer. Because these mixers tend to have rather poor LO-RF and RF-LO isolation, these tuned filters will help improve the port isolation by preventing the LO from appearing in the RF output, and the RF from being fed to the output of the LO source.

In many practical cases, the LO filter may be eliminated because it is difficult to make a filter that will track a variable LO frequency. In some cases, the receiver designer will use an untuned bandpass filter, while in others the output of the LO is applied directly to the source of the JFET through either a coupling capacitor or an untuned RF transformers.

The output of the unbalanced mixer contains the full spectrum of $mF_{RF} \pm nF_{LO}$ products, so a tuned filter is needed here also. The drain terminal of the JFET is the IF port in this circuit. The usual case is to use either a double-tuned L-C transformer (T1) as in *Figure 3-16*, or some other sort of filter. Typical non-LC filters used in receivers include ceramic and quartz crystal filters, and mechanical filters.

A MOSFET version of the same type of circuit is shown in *Figure 3-17*. In this circuit, a dual-gate MOSFET (e.g., 40673) is the active element. The RF is applied to gate-1, and the LO is applied to gate-2, with the LO signal level being sufficient to drive Q1 into nonlinear operation. A resistor voltage divider (R3/R4) is used to provide a DC bias level to gate-2. The source terminal is bypassed to ground for RF, and is the common terminal for the mixer.

In this particular case the LO input is broadband, and is coupled to the LO source through a capacitor (C3). The RF input, on the other hand, is tuned by a resonant bandpass filter (L1B/C1).

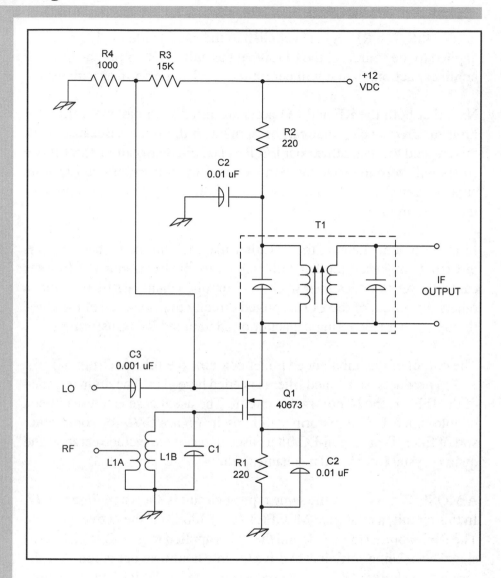

Figure 3-17.
Single-ended
MOSFET mixer.

Balanced Active Mixers

There are a number of balanced active mixers that can be selected. Many of these forms are now available in integrated circuit (IC) form. Because of the intense activity being seen in the development of telecommunications equipment (cellular, PCS and other types), there is a lot of IC development being done in this arena.

One of the earliest types of RF IC on the market was the differential amplifier. *Figure 3-18* shows the use of one of these ICs is as a mixer stage. Two transistors (Q1 and Q2) are differentially connected by having their emitter terminals connected together to a common-current source (Q3). The RF signal is applied to the bases of Q1 and Q2 differentially through transformer T1. The LO signal is used to drive the base terminal of the current source transistor (Q3). The collectors of Q1 and Q2 are differentially connected through a second transformer, T2, which forms the IF port.

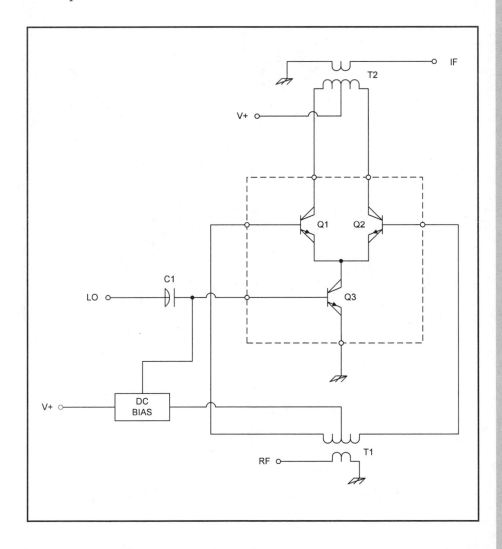

Figure 3-18.
Balanced NPN
differential amplifier
mixer.

Figure 3-19 shows one rendition of the Sabin (1970) double-balanced mixer used in some HF receivers. It offers a noise figure of about 3 dB. The Sabin mixer features a push-pull pair of high pinchoff voltage JFETs (Q1 and Q2) connected in a common-source configuration. The LO signal is applied to the common source in a manner similar to *Figure 3-17*.

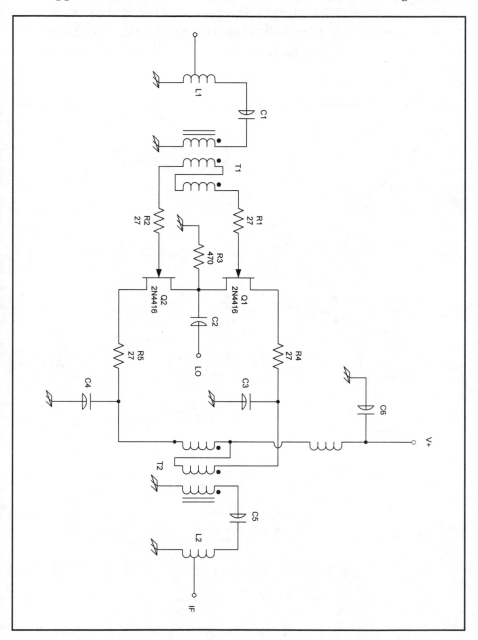

Figure 3-19.
Balanced MOSFET
mixer.

The gate circuits of Q1 and Q2 are driven from a balanced transformer, T1. This transformer is trifilar wound, usually on either a toroid or binocular BALUN core. The dots on the transformer windings indicate the phase sense of the winding. Note that the gate of Q1 is fed from a dotted winding end, while that of Q2 is fed from a non-dotted end. This arrangement ensures that the signals will be 180 degrees out of phase, resulting in the required push-pull action. Some input filtering, and matching the 1.5-kohm JFET input impedance to a 50- or 75-ohm system impedance (as needed), is provided by L1/C1.

The IF output is similar to the RF input. A second trifilar transformer (T2) is connected such that one drain is to a dotted winding end and other is to a non-dotted end of T2. Compare the sense of the windings of T1 and T2 in order to avoid signal cancellation due to phasing problems. IF filtering and impedance matching is provided by C5/L2. The tap on L2 is adjusted to match the 5.5-kohm impedance of the JFETs to system impedance.

A MOSFET double-balanced active mixer is shown in *Figure 3-20*. This circuit was discussed in DeMaw and Williams (1981). The dual-gate mixer is ideally suited to this type of application, but one must be cautious regarding selection. The 40673 device has the advantage of being relatively insensitive to casual electrostatic damage during handling, and is low in cost. It has sufficient gain in a circuit such as this to provide 15 to 20 dB of conversion gain. However, this gain is not without cost, as those devices are relatively easy to overdrive in the presence of large RF input signals. When this occurs the advantages of the device evaporate in increased IM products and degraded noise performance. The same circuit using 2N211 devices, or their equivalent, will produce less conversion gain (e.g., about -5 dB), but better overall performance. With an LO injection of 8 volts peak-to-peak, and a 10 dBm RF input signal, this circuit will exhibit a respectable third-order intercept point of +17 dBm.

An active double-balanced mixer based on NPN bipolar transistors is shown in *Figure 3-21* (Rhode 1994b). This circuit is usable to frequencies around 500 MHz. Normally, the use of non-IC transistors in a circuit such as this requires matching of the transistors for best perfor-

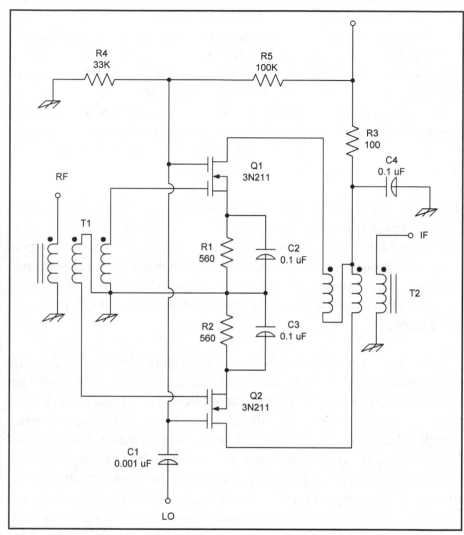

Figure 3-20.
High TOIP bipolar
balanced mixer.

mance. That need is overcome by using a bit of degenerative feedback for Q1 and Q2 in the form of unbypassed emitter resistors (R3 and R4).

The base circuits are driven with the LO signal from a BALUN transformer (T1) in a manner similar to the earlier JFET circuit. The output transformer, however, is rather interesting. It consists of four windings, correctly phased, with the IF being taken from the junction of two of the windings. The RF signal is applied to the remaining two windings of the transformer. This mixer exhibits a third-order intercept of +33 dBm, with a conversion loss of 6 dB, and only 15 to 17 dBm of LO drive power.

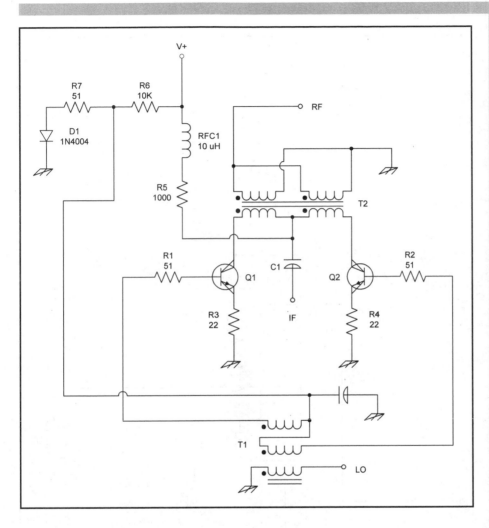

Figure 3-21.
Quad JFET
balanced mixer.

Although the use of bipolar transistors can result in an active double-balanced mixer with a high TOIP, there is a distinct trend today towards the use of JFET and MOSFET devices. Typical designs use four active devices. This approach is made easier by the fact that many IC makers are producing RF MOSFET and JFET products that include four matched devices in the same package.

Figure 3-22 shows a mixer circuit based on the use of four JFET devices (Q1-Q4). These transistors are arrayed such that the source terminals of Q1-Q2 are tied together, as are the source terminals of Q3-Q4. These source-pair terminals receive the RF input signal from transformer T2.

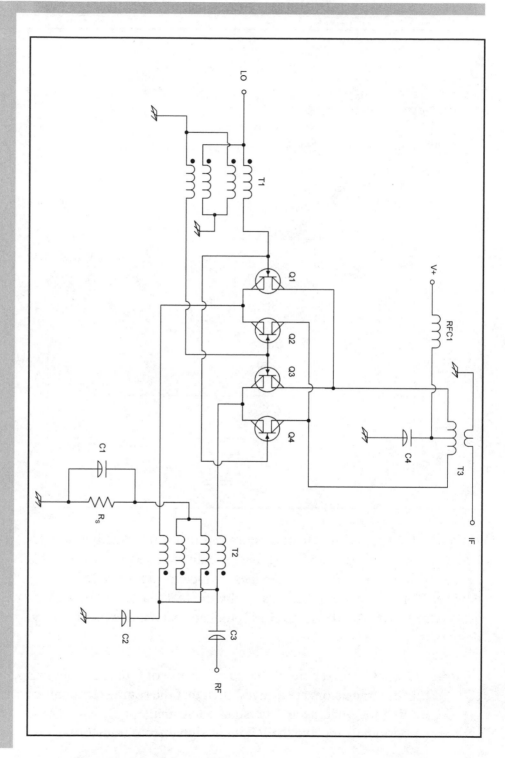

Figure 3-22.
Siliconix quad
MOSFET device.

The gates of these transistors are connected such that Q1-Q4 and Q2-Q3 are paired. The LO signal is applied differentially to these gates through transformer T1. The Q1-Q3 and Q2-Q4 drains are tied to the IF output transformer T3.

There are several quad FET ICs on the market that have found favor as mixers in radio receivers. The Siliconix SD5000 DMOS Quad FET is shown in *Figure 3-23*. This device contains four DMOS MOSFETs that can be used independently. When connected as a ring, they will form a mixer. Calogic carries the theme a little further in their SD8901 DMOS quad FET mixer IC (*Figure 3-24*). The FETs (Q1-Q4) are connected in a ring such that opposite gates are connected together to form two LO ports (LO1 and LO2). The RF signals are applied differentially across drain-source nodes Q1-Q2 and Q3-Q4. Similarly, the IF output is taken from the opposite pair of nodes: Q1-Q4 and Q2-Q3. The SD8901 comes in an eight-pin metal can package.

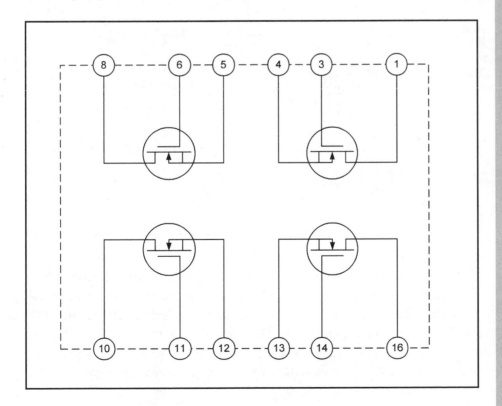

Figure 3-23.
Calogic DM8901
mixer.

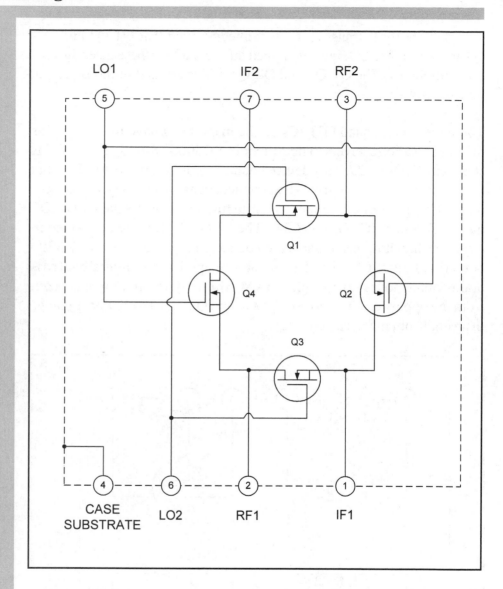

Figure 3-24.
DM8901 mixer
circuit.

The circuit for using the SD8901 (*Figure 3-25*) is representative of this class of mixers. The RF and IF output terminals are through transformers T1 and T2, respectively. The LO signal is applied directly to the LO1 and LO2 ports, but requires a J-K flip-flop divide-by-2 circuit. Note that this makes the LO signal a square wave rather than a sine wave. An implication of this circuit is that the LO injection frequency must be twice the expected LO frequency.

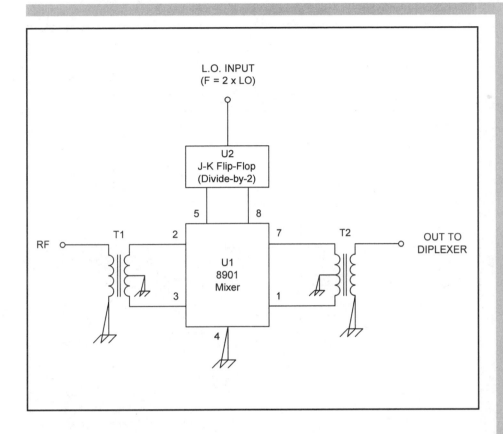

Figure 3-25.
Gilbert transcon-
ductance cell
circuit.

Gilbert Cell Mixers

The Gilbert transconductance cell (*Figure 3-26*) is the basis for a number of IC mixer (e.g., the NE-602 shown in *Figure 3-27*) and analog multiplier (e.g., LM-1496) devices. The circuit consists of two cross-connected NPN pairs fed from a common-current source. The RF signal is differentially applied to the transistors that control the apportioning of the current source between the two differential pairs. The LO signal is used to drive the base connections of the differential pairs.

IC Gilbert cell devices such as the Phillips/Signetics NE-602 (*Figure 3-27*) are used extensively in low-cost radio receivers. The Gilbert cell is capable of operating to 500 MHz. An on-board oscillator can be used to 200 MHz. One problem seen on such devices is that they often trade off dynamic range for higher sensitivity.

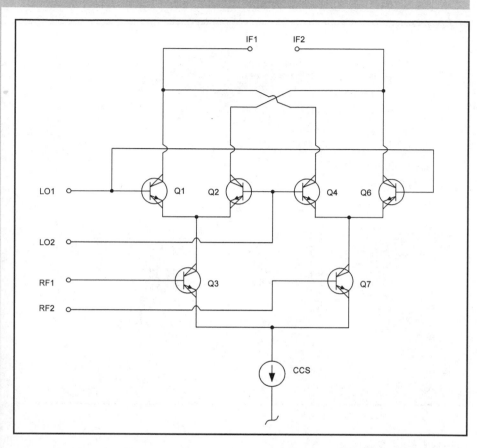

Figure 3-26.
Passive double
balanced mixer.

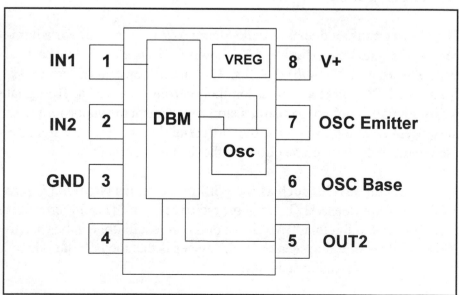

Figure 3-27.
NE-602 pinouts.

Passive Double-Balanced Mixers

The diode double-balanced mixer (*Figure 3-28*) is one of the more popular approaches to DBM design. It has the obvious advantage over active mixers of not requiring a DC power source. This circuit uses a diode ring (D1-D4) to perform the switching action. In the circuit shown in *Figure 3-28* only one diode per arm is shown, but some commercial DBMs use two or more diodes per arm. It is capable of 30 to 60 dB of port-to-port isolation, and is easy to use in practical applications.

With proper design, it is easy to build passive diode DBMs with frequency responses from 1 to 500 MHz, although commercial models are easily obtained into the microwave region. The IF outputs of the typical DBM can be DC to about 500 MHz.

The diodes used in the ring can be ordinary silicon small-signal diodes such as 1N914 and 1N4148, but these are not as good as hot carrier Schottky diodes (e.g., 1N5820, 1N5821, 1N5822, etc.). Whichever diodes are selected, however, they should be matched for use in the circuit because diode differences can deteriorate mixer performance. The usual approach is to match the diode forward voltage drop at some specified standard current such as 5 to 10 mA, depending on the normal forward current rating of the diode. Also of importance is matching the junction capacitance of the diodes.

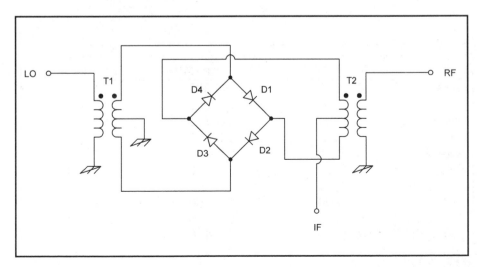

Figure 3-28.
Mini-Circuits
SRA/SBL DBM.

The DBM in *Figure 3-28* uses two BALUN transformers, T1 and T2, to couple to the diode ring. The double-balanced nature of this circuit depends on these transformers, and as a result the LO and RF components are suppressed in the IF output.

Diode DBMs are characterized according to their drive level requirements, which is a function of the number of diodes in each arm of the ring. Typical values of drive required for proper mixing action are 0 dBm, +3 dBm, +7 dBm, +10 dBm, +13 dBm, +17 dBm, +23 dBm and +27 dBm.

Figure 3-29 shows the internal circuitry for a commercially available passive DBM made by Mini-Circuits (Box 166, Brooklyn, N.Y., 11235, USA:Phone 714-934-4500; Web site http://www.minicircuits.com). These type no. SBL-x and SRA-x devices are available in a number of different characteristics (see *Table 3-1*).

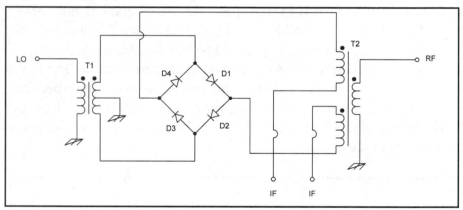

Figure 3-29.
Mini-Circuits DBM
package.

Table 3-1.
SBL-x and SRA-x
devices.

Type	LO Range (MHz)	IF Range (MHz)	Mid-Band Loss (dB)
SRA-1	0.5-500	DC-500	5.5-7.0
SRA-1TX	0.5-500	DC-500	5.5-7.0
SRA-1W	1-750	DC-750	5.5-7.5
SRA-1-1	0.1-500	DC-500	5.5-7.5
SRA-2	1-1000	DC-500	5.5-7.5
SBL-1	1-500	DC-500	5.5-7.0
SBL-1X	10-1000	5-500	5.5-7.5
SBL-1Z	10-1000	DC-500	5.5-7.5
SBL-1-1	0.1-400	DC-400	5.5-7.0
SBL-3	0.025-200	DC-200	5.5-7.5

The standard package for SRA/SBL devices is shown in *Figure 3-30*. The pins are symmetrical on 5.06 mm (0.20-inch) centers. Pin no. 1 is indicated by a blue dot insulator (the other insulated pins have green insulation). The noninsulated pins are grounded to the case. Pinouts for common SRA/SBL devices are shown in *Table 3-2*.

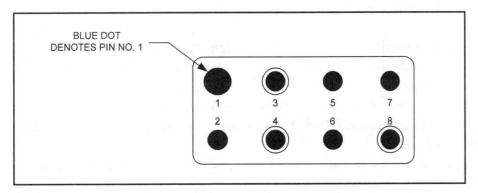

BLUE DOT
DENOTES PIN NO. 1

Figure 3-30.
Standard SRA/SBL
package

Type No.	LO	RF	IF	GND	CASE
SRA-1	8	1	3,4	2,5,6,7	2
SRA-1TX	8	1	3,4	2,5,6,7	2
SRA-1W	8	1	3,4	2,5,6,7	2,5,6,7
SRA-1-1	8	1	3,4	2,5,6,7	2
SRA-2	8	3,4	1	2,5,6,7	2,5,6,7
SBL-1	8	1	3,4	2,5,6,7	2
SBL-1X	8	1	3,4	2,5,6,7	2
SBL-1Z	8	1	3,4	2,5,6,7	XX
SBL-1-1	8	1	3,4	2,5,6,7	XX
SBL-3	8	1	3,4	2,5,6,7	XX

Table 3-2.
SRA/SBL pinouts.

The regular SRA/SBL devices use an LO drive level of +7 dBm, and can accommodate RF input levels up to +1 dBm. The devices will work at lower LO drive levels, but performance deteriorates rapidly, so it is not recommended.

Note that the IF output port is split into two pins (3 and 4). Some models tie the ports together, but for others an external connection must be provided for the device to work.

The nice thing about this type of commercially available mixer is that the system impedances are already set to 50 ohms. Otherwise, impedance matching would be necessary for them to be used in typical RF circuits. Note, however, that if a circuit or system impedance is other than 50 ohms, then a mismatch loss will be seen unless steps are taken to effect an impedance match.

The mismatch problem becomes considerably greater when the mismatch occurs at the IF port of the mixer. The mixer works properly only when it is connected to a matched resistive load. Reactive loads and mismatched resistive loads deteriorate performance. *Figure 3-31* shows a circuit using a passive diode DBM. The diplexer is a critical component to this type of circuit.

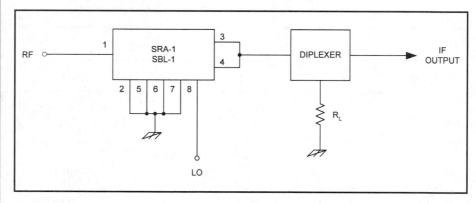

Figure 3-31.
Terminating a
passive DBM.

Diplexers

The diplexer is a passive RF circuit that provides frequency selectivity at the output, while looking like a constant resistive impedance at its input terminal. *Figure 3-32* shows a generalization of the diplexer. It consists of a *high-pass filter* and a *low-pass filter* that share a common input line, and are balanced to present a constant input impedance. With appropriate design, the diplexer will not exhibit any reactance reflected back to the input terminal (which eliminates the reflections and VSWR problem). Yet, at the same time it will separate the high- and low-frequency components into two separate signal channels. The idea is to forward the desired frequency to the output and absorb the unwanted frequency in a dummy load.

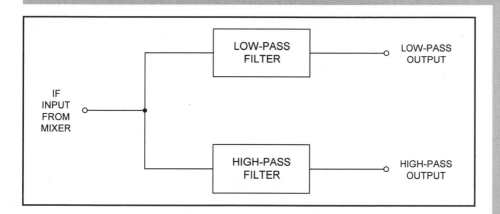

Figure 3-32.
Diplexer block
diagram.

Figure 3-33 shows the two cases. In each case, a mixer nonlinearly combines two frequencies, F1 and F2, to produce an output spectrum of mF1 ± nF2, where m and n are integers representing the fundamental and harmonics of F1 and F2. In some cases, we are interested only in the difference frequency, so will want to use the low-pass output (LPO) of the diplexer (*Figure 3-33A*). The high-pass output (HPO) is terminated in a matched load so that signal transmitted through the high-pass filter is fully absorbed in the load.

The exact opposite situation is shown in *Figure 3-33B*. Here we are interested in the sum frequency, so use the HPO port of the diplexer, and terminate the LPO port in a resistive load. In this case, the signal passed through the low-pass filter section will be absorbed by the load.

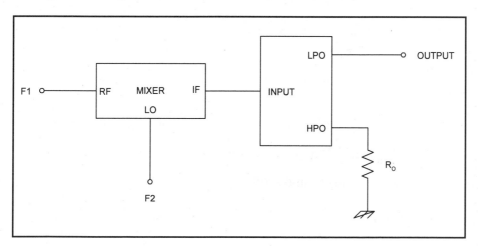

Figure 3-33A.
Low-pass diplexer.

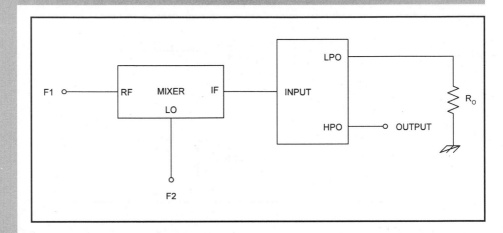

Figure 3-33B.
High-pass diplexer.

Bandpass Diplexers

Figures 3-34 and *3-35* show two different bandpass diplexer circuits commonly used at the outputs of mixers. These circuits use a bandpass filter approach, rather than two separate filters. In *Figure 3-34* a π-network approach, while the version in *Figure 3-35* is an L-network. In both cases,

$$Q = \frac{f_o}{BW_{3dB}} \qquad (3\text{-}7)$$

and,

$$\omega = 2\pi f_o \qquad (3\text{-}8)$$

Where:

f_o is the center frequency of the passband in hertz (Hz)

BW_{3dB} is the desired bandwidth in hertz (Hz)

Q is the relative bandwidth

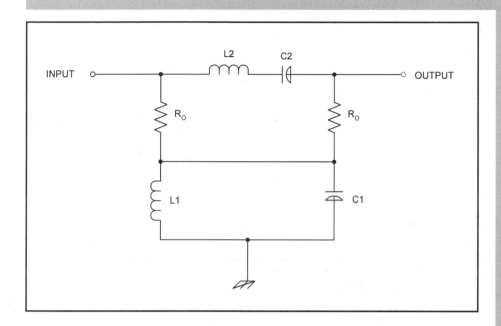

Figure 3-34.
Pi-diplexer.

For the circuit of *Figure 3-34*:

$$L2 \ = \ \frac{R_o \, Q}{\omega} \qquad\qquad (3\text{-}9)$$

$$L1 \ = \ \frac{R_o}{\omega \, Q} \qquad\qquad (3\text{-}10)$$

$$C2 \ = \ \frac{1}{R_o \, Q \, \omega} \qquad\qquad (3\text{-}11)$$

$$C1 \ = \ \frac{Q}{\omega \, R_o} \qquad\qquad (3\text{-}12)$$

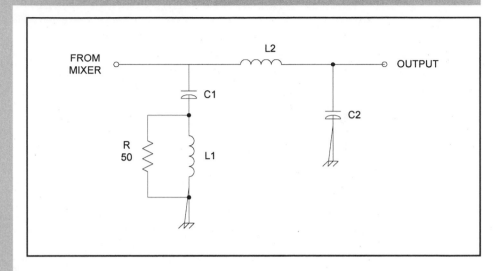

Figure 3-35.
L-section diplexer.

For the circuit of *Figure 3-35*:

$$L2 = \frac{R_o\, Q}{\omega} \qquad\qquad (3\text{-}13)$$

$$L1 = \frac{R_o}{\omega\, Q} \qquad\qquad (3\text{-}14)$$

$$C1 = \frac{1}{L1\, \omega^2} \qquad\qquad (3\text{-}15)$$

$$C2 = \frac{1}{L2\, \omega^2} \qquad\qquad (3\text{-}16)$$

Double DBM

The normal passive diode DBM provides relatively high TOIP and
-1 dB compression points. It also provides a high degree of port-to-port
isolation because the switching action of the diodes in the ring are shut
off at the instances where they would feed through the other ports. But

where an even higher degree of performance is needed designers sometimes opt for the double DBM as shown in *Figure 3-36*. The -1 dB compression point is usually ≤ 4 dB.

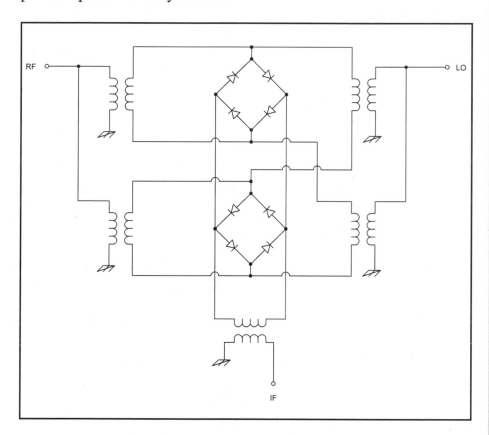

Figure 3-36.
Double Double-
balanced mixer.

Image Reject Mixers

In cases where very good image rejection performance is needed in a receiver, a circuit such as *Figure 3-37* can be used. This circuit uses a pair of passive DBMs, a 0-degree power splitter and two 90-degree power splitters to form an image reject mixer. The LO ports of Mixer-1 and Mixer-2 are driven in-phase from a master LO source. The RF input, however, is divided into quadrature signals and applied to the respective RF inputs of the two mixers. The IF outputs of the mixers are then re-combined in another quadrature splitter, to form separate USB and LSB IF outputs.

VHF/UHF Microwave Mixer Circuits

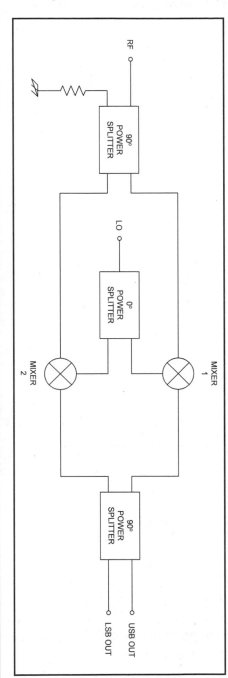

Figure 3-37.
Image reject mixer.

When the frequencies used for LO, RF and IF begin to reach into the VHF and above region, design approaches change a bit. *Figure 3-38* shows a simple single-diode unbalanced mixer. Variants of this circuit have been used in UHF television and other types of receivers. The circuit is enclosed in a shielded space in which a strip inductor (L1) and a variable capacitor (C1) form a resonant circuit. The LO and RF signals are applied to the mixer through coupling loops to L1. A UHF signal diode is connected to L1 at a point that matches its impedance. External to the mixer chamber, an IF filter is used to select the mixer product desired for the IF section of the particular receiver.

A single-balanced mixer is shown in *Figure 3-39*. This mixer circuit uses two diodes, D1 and D2, connected together and also to the ends of a half-wavelength 100-ohm transmission line. The LO signal is applied to D2 and one end of the transmission line. IF and RF filters are used to couple to the RF and IF ports.

These mixers suffer from RF and LO components appearing in the output. *Figure 3-40* shows an improved ver-

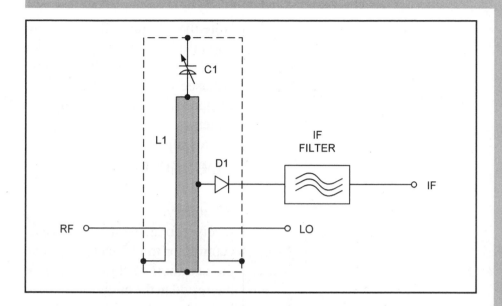

Figure 3-38.
VHF/UHF single-
ended diode mixer.

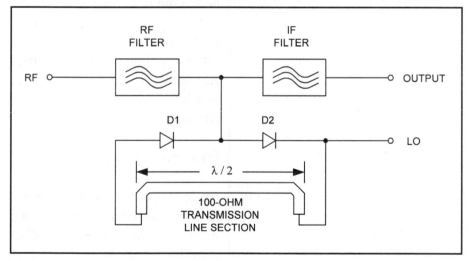

Figure 3-39.
UHF single-
balanced mixer.

sion that will solve the problem. It is used in the UHF and microwave regions. The LO and RF input signals are applied to two separate ports of a quadrature hybrid coupler. The input and output filtering are made using printed-circuit board transmission lines. Each of these are quarter wavelength, although the actual physical lengths must be shortened by the velocity factor of the printed-circuit board being used. A printed-circuit RF choke (RFC) is used to provide a return connection for the diodes.

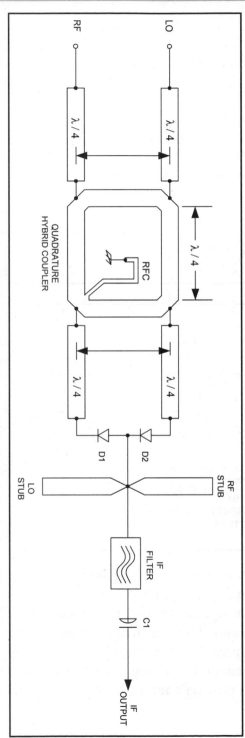

Figure 3-40.
Microwave mixer.

Note the RF and LO stubs at the output of the mixer, prior to the input of the IF filter. These stubs are used to suppress RL and LO components that pass through the mixer.

Special Circuits

Whether you are building a radio receiver for general shortwave reception, or the IF section of a radio astronomy setup, you will be interested in the circuits discussed here. One is a "universal" IF amplifier that can be used at any of the standard frequencies, and the other is a logarithmic amplifier. But more of that later. The IF amplifier can also be used for an RF amplifier at some frequencies.

Universal IF Amplifier

An IF amplifier is used in superheterodyne receivers to perform most of the amplification and selectivity functions. Typically, IF gains run in the 60 to 90 dB range, depending on the receiver design. The IF amplifier in *Figure 3-41* is based on the popular MC-1350P integrated circuit. This chip is easily available through any of the major mail-order parts houses, and many small ones. It has been popularized in amateur radio technical articles aimed at the QRP (low-power) crowd by Doug DeMaw's books and articles. It is basically a variation on the LM-1490 type of circuit, but is a little easier to apply in my opinion.

If you have difficulty locating MC-1350P devices, the exact same chip is available in the service replacement lines, such as ECG and NTE. These parts lines are sold at local electronics parts distributors, and are intended for the service repair shop trade. I used actual MC-1350P chips in one version, and NTE-746 (same as ECG-746) chips in the other, without any difference in performance. The NTE and ECG chips are actually purchased from the sources of the original devices, and then renumbered.

The circuit is shown in *Figure 3-41*. Two MC-1350P devices in cascade are used. Each device has a differential input (pins 4 and 6). These pins are connected to the link windings on IF transformers (e.g., T2 at device U1 and T3 at U2). In both cases, one of the input pins are grounded for AC (i.e., RF and IF) signals through a bypass capacitor (C2 and C4).

In the past I've had difficulties applying the MC-1350P devices when two were used in cascade. The problem is that these are high-gain chips, and any coupling at all will cause oscillation. I've built several real good MC-1350P oscillators...the problem is that I was building IF *amplifiers*. The problem is basically solved by two tactics that I'd ignored in the past. This time I reversed the connections to the input terminals on the two devices. Note that pin #4 is bypassed to ground on U1, while on U2 it is pin #6. The other tactic is to use different-value resistors at pin #5.

Pin #5 on the MC-1350P device is the gain control pin. It is used to provide either *manual gain control* (MGC) or *automatic gain control* (AGC). Voltage applied to this pin should be between +3 and +9 volts, with the highest gain being at +3 volts and nearly zero gain at +9 volts.

The outputs of the MC-1350P are connected to the primaries of T3 and T4. Each output circuit has a resistor (R2 and R5) across the transformer winding. The transformers used are standard "transistor radio" IF transformers. I used the Mouser 42IF123 (10.7 MHz) devices. How-

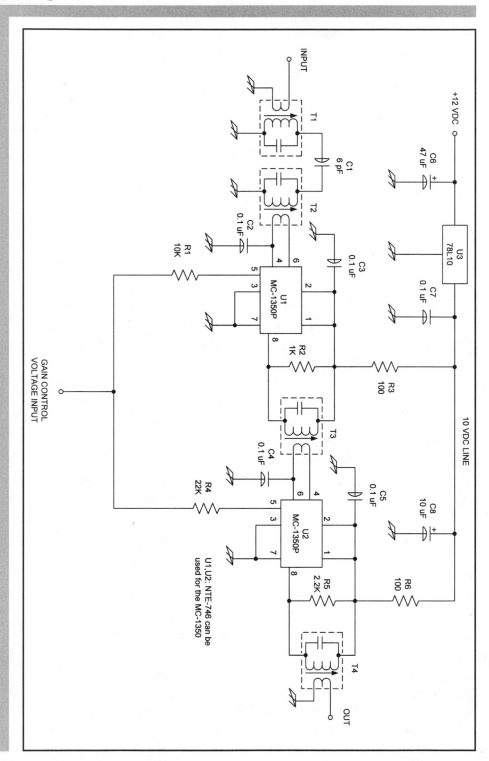

Figure 3-41.
Two-stage IF
amplifier based on
MC-1350P chips.

ever, almost any other IF transformer can be used as well. In one version I used Toko 455 kHz IF transformers purchased from Digi-Key, and in another 455 kHz transformers of unknown origin from another source.

The DC power is applied to the MC-1350P devices though pins 1 and 2, which are connected together. Bypass capacitors C3 and C5 are used to decouple the DC power lines, and thereby prevent oscillation. All of the bypass capacitors (C2, C3, C4 and C5) should be mounted as close to the bodies of U1 and U2 as possible. They can be disk ceramic devices, or some of the newer dielectric capacitors, provided of course that they are rated for operation at the frequency you select. Most capacitors will work to 10.7 MHz, but if you go to 50 MHz or so, some capacitor types might show too much reactance (disk ceramics work fine at those frequencies, however).

The DC power supply should be regulated at some voltage between +9 and +12 VDC. More gain can be obtained at +12 VDC, but I used +10 VDC with good results. In each power line there is a 100 ohm resistor (R3 and R6) which helps provide some isolation between the two devices. Feedback via the power line is one source of oscillation in high-frequency circuits.

The DC power line is decoupled by two capacitors. C7 is a 0.1 µF capacitor, and is used to decouple high frequencies that either get in through the regulator or try to couple from chip to chip via the DC power line (which is why they are called "decoupling" capacitors). The other capacitor (C8) is a 10 to 100 µF device used to smooth out any variations in the DC power or to decouple low frequencies that the 0.1 µF doesn't take out effectively.

The RF/IF input circuit deserves some comment. I elected to use a double-tuned arrangement. This type of circuit is of a category that are coupled via a mutual reactance. Various versions of this type of circuit are known (see my book *Secrets of RF Circuit Design* - available through Amazon Books), but I elected to use the version that uses a capacitive reactance (C1) at the "hot" end of the L-C tank circuits. Because I am using standard IF transformers, I coupled the tuned sides through the capacitor.

Coupling in and out of the network is provided by the transformer coupling links.

Construction of the circuit is made easier if you use PCB.

Figure 3-42 shows two views of the completed project. The view in *Figure 3-42A* is the printed-circuit board with components assembled. (Note: This photo was taken before C8 was installed.) The box shown in *Figure 3-42B* is one of the SB "Lab" series made by SESCOM, Inc. This company makes a number of really interesting boxes for RF and other builders. I've bought quite a number of them because they are "tighter" for RF than some of the other aluminum and steel construction boxes on the market...and that means better shielding (which is all-important in RF work).

SESCOM, Inc.
2100 Ward Drive,
Henderson, NV,
89015-4249

The connectors for input and output can be any coaxial connector. I used RCA phono connectors because of space constraints in the project that this IF amplifier will be used in. At 10.7 MHz the loss with RCA phono connectors (which are normally used at audio) is negligible, so ought not pose any problem. At higher frequencies I recommend BNC connectors, SMC connectors, or SO-239 "UHF" connectors (if you've got lots of room!).

The power and gain control connections are bought through the aluminum box wall through 1000 pF feedthrough capacitors (also in the SESCOM catalog, plus other sources). Two kinds are available, one solder-in and the other screw thread mounted. For aluminum boxes the

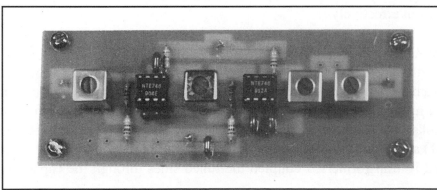

Figure3-42A.
IF amp circuit board
with installed
components.

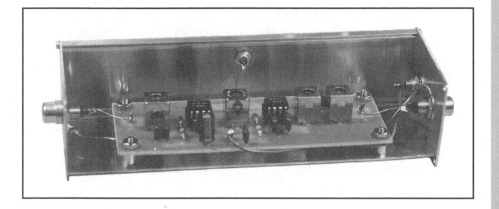

Figure 3-42B.
Completed project.

screw thread is needed because it is difficult to solder to aluminum. Both types are available in either 1000-pF or 2000-pF values, either of which can be used in this application. If you elect to use some other form of connector, then add disk ceramic capacitors (0.001 μF) to the connector, right across the pins as close as possible to the connector.

There are several ways to make this circuit work at other frequencies. If you want to use a standard IF frequency up to 45 MHz or so, then select one with the configuration shown in *Figure 3-41* (these are standard).

If you want to make the circuit operate in the HF band on a frequency other than 10.7 MHz, then it's possible to use the 10.7 MHz transformer. If the desired frequency is less than 10.7 MHz, then add a small-value fixed or trimmer capacitor in parallel with the tuned winding. It will add to the built-in capacitance, reducing the resonant frequency. I don't know how low you can go, but I've had good results at the 40-meter amateur band (7-7.3 MHz) using additional capacitance across a 10.7 MHz IF transformer.

On frequencies higher than 10.7 MHz you must take some more drastic action. Take one of the transformers and turn it over so that you can see the pins. In the middle of the bottom header, between the two rows of pins, there will be an indentation containing the tuning capacitor. It is a small tubular ceramic capacitor. (You may need a magnifying glass to see it well if your eyes are like mine.) If it is color coded, then you can obtain the value using your knowledge of the standard color code. Take a small screwdriver and crush the capacitor. Yeah, you read it right: *crush the capacitor*. Clean out all of the debris to prevent shorts at a later time. You now have an untuned transformer with an inductance right around 2 µH (I measured values between 1.8 and 2.6 µH using the inductance function of my DMM). Using this information, you can calculate the required capacitance using:

$$C = \frac{2.53 \times 10^{10}}{f^2 \, L_{\mu H}} \text{ picofarads} \qquad (3\text{-}17)$$

Where:

C is the capacitance in picofarads

f is the desired frequency in kilohertz

$Lm_{\mu H}$ is the inductance in microhenrys

Equation 3-17 is based on the standard L-C resonance equation, solved for C, and with all constants and conversion factors rolled into the numerator. If you know the capacitance that must be used, and need to calculate the inductance, then swap the L and C terms in Equation 3-17.

If the original capacitor was marked as to value or color coded, then you can calculate the approximate capacitance needed by taking the ratio of the old frequency to the new frequency, and then squaring it. The square of the frequency ratio is the capacitance ratio, so multiply the old capacitance by the square of the frequency ratio to find the new value. For example, suppose a 110 pF capacitor is used for 10.7 MHz, and you want to make a 20.5 MHz coil. The ratio is (10.7 MHz/20.5 MHz)² = 0.272. The new capacitance will be about 0.272 x 110 pF = 30 pF. For other frequencies, you might consider using homebrew toroid inductors.

A variation on the theme is to make the circuit wideband. This can be done for a wide portion of the HF spectrum by removing the capacitors from the transformers, and not replacing them with some other capacitor.

In the example above I used a frequency of 20.5 MHz (i.e., 20,500 kHz) to make the calculation. The reason is simple: it is a prime RadioScience Observing frequency for finding natural radio signals from Jupiter (see my book *RadioScience Observing, Volume 1*). It is possible to build a tuned radio frequency (TRF) receiver for that frequency using this circuit, but retuned to 20.5 MHz.

Logarithmic RF/IF Amplifier

A logarithmic amplifier produces an output signal that is proportional to the logarithm of the input signal. There are a number of applications for these amplifiers. Some of them take advantage of the range compression that takes place, making it easier to get a large dynamic range. In other cases, log amps are used to convert linear signal levels to the logarithmic decibel (dB) scale. The AD8307 log amp (*Figure 3-43*) will fit a lot of these uses, including RadioScience Observing receivers. It has a 92 dB dynamic range from -75 dBm to +17 dBm, and operates over the range DC to 500 MHz with ± 1 dB linearity. It produces a slope of 25 mV/dB and an intercept of -84 dBm, but this can be trimmed to other values using the SLOPE ADJ and INTERCEPT ADJ potentiometers shown in *Figure 3-43*. The DC power requirements are +2.7 to 5 VDC at 7.5 mA.

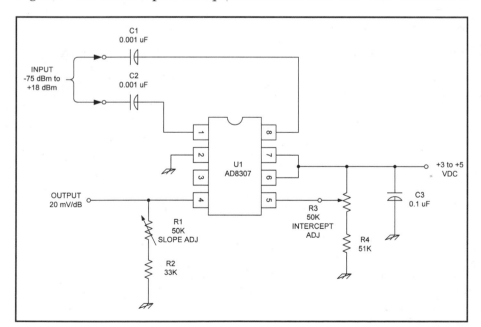

Figure 3-43. Logarithmic amplifier.

If you want to see the entire spec sheet for the AD8307, then go to the Analog Devices web site (http://www.analog.com). The data sheet is available in *.PDF format, so you will need Adobe *Acrobat Reader* to display and print it. The device can be purchased for around $18 from Newark Electronics and other Analog Devices distributors.

References

Carr, Joseph J. (1997). *Microwave and Wireless Communications Technology*. Boston: Newnes.

Carr, Joseph J. (1996). *Secrets of RF Circuit Design 2nd Edition*. New York: McGraw-Hill.

DeMaw, D. and G. Collins (1981). "Modern Receiver Mixers for High Dynamic Range". *QST*, May 1981; Newington, CT, USA: ARRL.

Gilbert, B. "Demystifying the Mixer," self-published monograph, 1994.

Hagen, Jon B. (1996). *Radio-Frequency Electronics: Circuits and Applications*. Cambridge (UK):Cambridge Univ. Press.

Hardy, James (1979). *High-Frequency Circuit Design*. Reston, VA: Reston Publishing Co. (Division of Prentice-Hall).

Hawker, P. ed, "Super-Linear HF Receiver Front Ends," Technical Topics, *Radio Communication,* Sep 1993, pp 54-56.

Hayward,W. *Introduction to Radio Frequency Design.* (Newington, CT: ARRL, 1994).

Horowitz P. and W. Hill, *The Art of Electronics, 2nd ed.* (New York: Cambridge University Press, 1989).

Joshi, S. "Taking the Mystery Out of Double-Balanced Mixers," *QST*, Dec 1993, pp 32-36.

Kinley, R. Harold (1985). *Standard Radio Communications Manual: With Instrumentation and Testing Techniques.* Englewood Cliffs, NJ: Prentice-Hall.

Laverghetta, Thomas S. (1984). *Practical Microwaves.* Indianapolis, IN: Howard W. Sams.

Liao, Samuel Y. (1990). *Microwave Devices & Circuits.* Englewood Cliffs, NJ: Prentice-Hall.

Makhinson, J. "High Dynamic Range MF/HF Receiver Front End," *QST*, Feb 1993, pp 23-28. Also see Feedback, *QST*, Jun 1993, p 73.

Rohde U., J. Whitaker, and T. Bucher, *Communications Receivers 2nd Ed.* (New York: McGraw-Hill Book Co, 1988).

Rohde U. and T. Bucher, *Communications Receivers: Principles and Design.* (New York: McGraw-Hill Book Co, 1988).

Rohde, U. (1994a) "Key Components of Modern Receiver Design," Part 1, *QST*, May 1994. Newington, CT, USA: ARRL; pp 29-32

Rohde, U. (1994b) "Key Components of Modern Receiver Design," Part 2, *QST*, Jun 1994. Newington, CT, USA: ARRL; pp 27-31

Rohde, U. (1994c) "Key Components of Modern Receiver Design," Part 3, *QST*, Jul 1994. Newington, CT, USA: ARRL; pp 42-45.

Rohde, U. "Testing and Calculating Intermodulation Distortion in Receivers," *QST*, Jul 1994. Newington, CT, USA: ARRL; pp 3-4.

Sabin, Williams (1970). "The Solid-State Receiver", *QST*, May 1970; Newington, CT, USA: ARRL; pp 35-43.

Sabin, William E. and Edgar O. Schoenike, editors (1998). *HF Radio Systems & Circuits 2nd Edition*. Atlanta: Noble Publishing.

Shrader, Robert L. (1975). *Electronic Communication 3rd Edition*. New York: McGraw-Hill.

Vizmuller, Peter (1995). *RF Design Guide*. Boston/London: Artech House.

Chapter 4
LIGHTNING OBSERVED

Chapter 4
Lightning Observed

Lightning is a familiar and often feared natural phenomenon. Although lightning is seen in all areas of the country, southern Arizona and central Florida are areas of extremely high *flash density* (i.e., the number of cloud-to-ground strikes per square mile per year). When lightning strikes, property can be damaged, fires started, trees can be split in two (or, oddly, the bark can come off leaving the tree core intact), and, sadly, sometimes people are killed. It's not surprising that the electrical power industry is a leader in lightning research because of the damage done to their power lines by lightning.

Types of Lightning

Cloud-to-ground lightning passes from a cloud overhead to the Earth beneath (*Figure 4-1*). Most cloud-to-ground strikes occur in the higher latitudes, although recent research indicates a more important variable is cloud top height. Cloud-to-ground lightning is what causes injury and damages. Cloud-to-cloud or "inter-cloud" lightning passes from one cloud to another. Intra-cloud lightning appears within a single cloud. The "jagged" lightning that we see so often is called *chain lightning*, while *sheet lightning* is a generalized bright flash.

Other types of lightning may or may not be little more than myths, poor observations (e.g., optical illusions), variations on other types, or real, depending on who you consult. These include ball lightning, tubular lightning, bead lightning, silent lightning, cloud-to-air lightning (as opposed to cloud-to-cloud), and heat lightning.

Lightning Facts

Human casualties are common. The daughter of a former governor of Virginia was struck at the National Guard Camp Pendleton (Virginia)

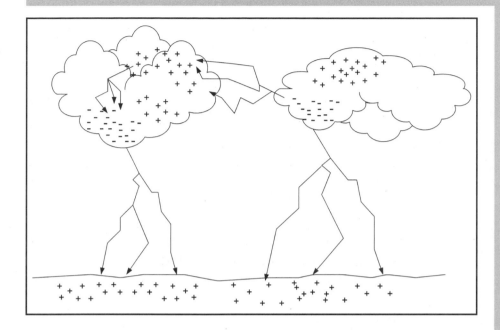

Figure 4-1.
Sources of
lightning strikes

beach in southeast Virginia. Florida saw 298 confirmed lightning deaths in the 30-year period 1959 to 1989. Between 1940 and 1989, there were 8,103 confirmed lightning deaths in the United States (SIRS 1990, *Earth Science*, Article 67). According to the National Oceanic and Atmospheric Administration (NOAA), lightning killed 88 people in 1995.

About 30 percent of the people struck each year are killed. I would have guessed that nearly all of them would have been killed, with only a few survivors, but the statistics apparently suggest otherwise. Those people who are killed usually succumb to cardiac arrest. The survivors frequently suffer serious heart problems thereafter, and most have to be treated for severe burns. One source claimed that there are cases on record where lightning did not even break the skin, but traveled the wet surface of their body to ground. The skin was said to be severely burned, similar to scalding.

It is estimated that Earth is struck by lightning 20 million times per year. If you believe the old myth about "lightning never strikes the same place twice," then consider the fact that the Empire State Building in New York City takes about twelve strikes per year. Radio and television antenna towers are also struck frequently, although I hasten to add that they do not "attract" lightning that would not have come anyway. (They tend to act as lightning rods.)

The lightning is generated up to 15,000,000 volts of electrical potential. According to one source, there is normally an atmospheric potential cloud-to-ground of 200,000 to 500,000 volts, and a constant but miniscule current of 10^{-12} amperes. The temperature inside the lightning bolt is 15,000 to 60,000 degrees centigrade, which is several times the temperature on the Sun. The lightning bolt travels at speeds up to 100,000,000 feet per second. A lightning bolt is actually a series of strokes, averaging about four. Duration varies from a few nanoseconds upward, but about thirty microseconds (30 μS) is said to be average. The average peak power per stroke is 10^{12} watts.

Many people struck by lightning were in unsafe situations; for example, on an open beach or in an open field such as a golf course. It is also not smart to be in a boat on the water. Standing next to a tree or other tall object (e.g., radio tower) is also rather dangerous, as the lightning may easily be attracted to that object. In one case, a group of golfers in Massachusetts were struck when lightning struck a nearby tree, then traveled underground to a covered pavilion where they had taken refuge from the storm. Carrying an umbrella with a metal tip above the canopy has caused lightning injuries, presumably because it looks to the lightning like a lightning rod...and you look like a ground wire!

As a general (but not absolute) rule, as long as you can hear thunder, the danger of strike is small. Lightning travels at the speed of light (3×10^8 meters/second), while sound travels at 331.3 meters per second at 0 °C (sound velocity changes a bit with air temperature), or 1,087 feet per second. As a result, the flash of light arrives instantaneously, while the sound of thunder arrives a short time later (measured in seconds).

A "rule of thumb" is to count the seconds between the flash and thunder clap, and then divide by five, to find the number of miles to the lightning bolt. One reason why I doubt validity of the "can hear thunder" advice is that I've been under thunderstorms where the lightning flash and thunder were very nearly simultaneous, indicating it was right overhead.

More precise measurements of distance can be done using a good stopwatch to measure the time of arrival of the thunderclap. From that data you can calculate the distance from $D = VT$, where D is the distance and V is the velocity (if you use m/s for T then D is in meters, but if ft/s is used T is in feet). You can further refine your measurements by consulting a table of sound velocity related to temperature and atmospheric pressures.

Some Lightning Detection History

Before we talk about early lightning research let me hasten to add a caution: **DON'T EVEN THINK ABOUT DOING THESE EXPERIMENTS YOURSELF**! People who've tried to duplicate these experiments have been killed in the attempt.

Electrical phenomenon was researched starting in the early eighteenth century. It was noticed that certain substances, when rubbed, produced static electricity arcs. It was also found that a type of capacitor called a Leyden Jar could store electrical charge. It was believed, correctly of course, that these sparks looked enough like lightning to suggest a connection.

In May 1752, Thomas Francois D'Alibard in France performed an experiment that Benjamin Franklin had failed at the year before. The experimenter stood on an electrical stand, holding an iron rod in one hand. The idea was to produce an electrical arc between that rod and another rod in the other hand connected to a grounded iron wire. In an attempt to repeat the experiment in July 1753, Swedish physicist G.W. Richmann, working in Russia, was killed by the lightning.

Benjamin Franklin conducted his famous kite experiment in 1752. He used a rain dampened kite string connected to a key at the bottom. The lightning traveled down the string, and jumped from the key to a dry silk

Most Common Activities Leading to Lightning Strikes on Humans (from NASA web site http:www.thunder.msfc. nasa.gov/primer.html)

1. Working, playing or walking in open fields.

2. Boating, fishing, swimming

3. Working on heavy farm or road construction equipment.

4. Playing golf (!!!)

5. Talking on the telephone.

6. Using electrical appliances

ribbon tied to Franklin's knuckles, and then through his grounded body. Others who attempted this experiment were killed. It's interesting to wonder what the American republic would look like if Franklin, if he'd been killed in 1752, was not alive to provide wisdom to the Constitutional Convention. (His compromise suggestion, which reconciled the "large states vs. small states" controversy, is supposedly how we got the bicameral Congress.)

When photography and spectroscopy was invented in the nineteenth century, additional facts were learned about lightning. Time-resolved photography was used to count the number of strokes per lightning strike. Current measurements were made by Pockels in Germany during 1897-1900. He analyzed the magnetic fields induced by the lightning currents to estimate the current level from basic electrical fundamentals. C.T.R. Wilson (c1920s), who also invented the cloud chamber, used electric field measurements to study lightning. Wilson hypothesized that the electrical fields cause such great ionization that it's possible for discharges to occur between the clouds and the upper atmosphere. Seeming confirmation of the 1925 theory was provided on 28 April 1990 when Shuttle mission STS-32 video taped a single luminous discharge in the stratosphere.

Modern lightning research still measures electrical and magnetic fields, but adds other techniques to their armamentarium. Cameras with electrically triggered shutters can be used with photosensors to make photographs of lightning strikes. High-speed movie cameras are also used, although the invention of charge-coupled device (CCD) sensors (which are used in video cameras) has allowed a number of new instruments to emerge.

Other researchers rely on radio waves, especially the whistlers, spherics and broadband RF noise generated in the ELF/VLF/LF portions of the radio spectrum. This type of research is also easily conducted by amateur scientists.

At a site in Florida, researchers fire three-foot high solid-fuel rockets into thunderclouds. The rockets trail a grounded wire that conducts the stroke to earth (**AMATEUR ROCKETEERS: DON'T DO THIS!**). The Space Shuttle and satellites are also being used for lightning research. Some hauntingly beautiful video clips of space shuttle lightning are available from NASA web sites. High-altitude aircraft lightning research revealed newly discovered phenomenon called sprites and jets.

Storm Scope

Lightning research can be terribly dangerous, especially if it puts you out in the open or if you are foolish enough to mimic professional methods such as ground-wire tethered rockets. There are, however, some instruments that are easily built and relatively safe to use. The storm scope is one such instrument.

Figure 4-2 shows the basic configuration. This project was originally published by Thomas P. Leary in the June 1964 *QST* magazine (although his implementation used vacuum tube amplifiers). The storm scope consists of a pair of orthogonal small loop antennas, one oriented north-south (N-S loop) and the other east-west (E-W loop). Keep in mind that a small loop antenna (e.g., < 0.18λ wire length) shows a figure-8 pattern

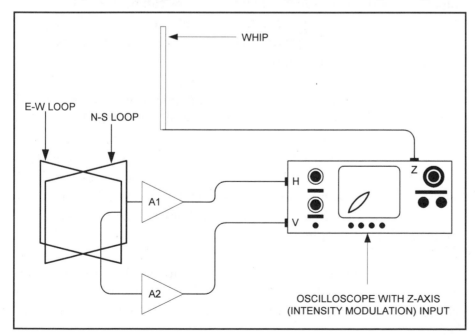

Figure 4-2.
Lightning detector
block diagram.

with nulls broadside to the loop plane, and maxima off the ends. Align the ends of the loop (maximum sensitivity direction) in E-W and N-S directions. A compass will make the accuracy better.

The loops are fed through differential amplifiers, with gains in the 20 to 80 dB range, to the vertical and horizontal plates of the oscilloscope. The reason for the wide gain scale is that oscilloscopes vary. The original project connected the amplifier outputs directly to the oscilloscope deflection plates. Modern 2-channel oscilloscopes can be used in the X-Y to form a vectorscope, of which the storm scope is a variation on the theme. Although any 2-channel scope with an X-Y mode is usable, a directional ambiguity exists unless the scope also has a Z-axis (a.k.a. intensity modulation) input. This input is often present, but hidden on the rear panel of the scope. The Z-axis us used to intensity modulate the screen with a signal from a sense whip or vertical antenna near the two-X loop. It may be necessary to amplify the Z-axis input signal, but in that case a single-ended rather than differential amplifier is used.

Figure 4-3 shows the loop and amplifier configuration. The loop consists of at least fifty turns (50-t.) of small gauge insulated wire in a three-foot (3-ft.) loop. Either square or circular loops can be used, although it appears to me that the square is more easily constructed. The loop is untuned. The output of the loop is applied to the input terminals of the differential or push-pull amplifier.

It is critical to shield the loops. Wrap either copper foil or aluminum foil (or tape) over the entire loop, except for a small quarter-inch or so gap along the top edge. (This prevents the shield from acting like a single-turn shorted loop itself.) The shielded loop will be less prone to pattern distortions from capacitive coupling. In addition, it responds largely to the magnetic component of the lightning electromagnetic field. It is less sensitive to electrical fields, so will not pick up as much locally generated power line and appliance noise.

I constructed a loop pair using 0.75" x 4" x 36" lumber. Although my own woodworking skills leave something to be desired (a polite way to say "atrocious"), others are able to make a better job of the method

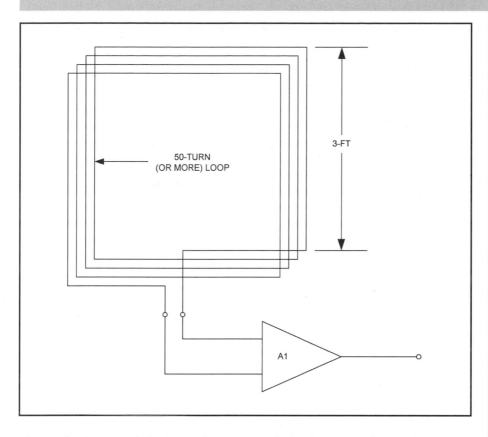

Figure 4-3.
Loop winding detail.

shown in *Figure 4-4*. A notch (*Figure 4-4A*) is cut at the center of the top and bottom members of each loop. The depth of the notch is the thickness of the lumber, while the width is sufficient to snug-fit the other piece of wood in it (*Figure 4-4B*).

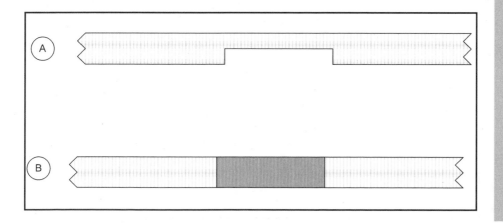

Figure 4-4.
Detail for making loop crosspieces.

The square loop will tend to "trapezoid" out of shape if left to its own devices. As a result, some or all of the methods of *Figure 4-5* are used at the four corners of each loop. In *Figure 4-5A* the wood elements are butted together, glued and nailed (or screw fastened). A small block, about 0.5 inch square and as wide as the loop arm, is glued in the corner to give strength. In *Figure 4-5B* the same method is used, but the joint between the wood members is a bit different. I am told this method is stronger, even though it is more difficult. Finally, a triangular gusset plate cut from a thin sheet of plywood or spruce modeling lumber is shown at *Figure 4-5C*.

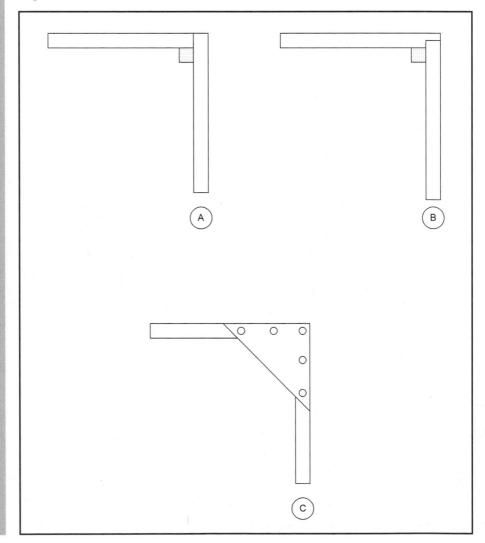

Figure 4-5.
Loop corner
mechanical details.

Figure 4-6 shows the wiring of the loop. I used 50-conductor ribbon cable, although one continuous loop of #26 (or so) enameled wire could be used as well (It takes more than 600 feet of wire per loop for this approach!). You can also use 60- or 64-conductor cable. The sensitivity of the loop is improved with more turns or a larger length per side. You can, for example, build a four- or five-foot loop, or use multiple runs of 50-conductor ribbon cable. In one case, a whistler/spheric hunter used 126 conductors made from intercom cable.

If you opt for the simpler ribbon cable method, then cross-connect adjacent turns so that one continuous loop is formed. This is a tedious chore, but it's made a lot easier if you use printed-circuit perf-board (or some of the project boards from *Radio Shack*) to make the connections. It is OK to lay the N-S and E-W loops over one another because their orthogonal geometry makes them minimally interactive.

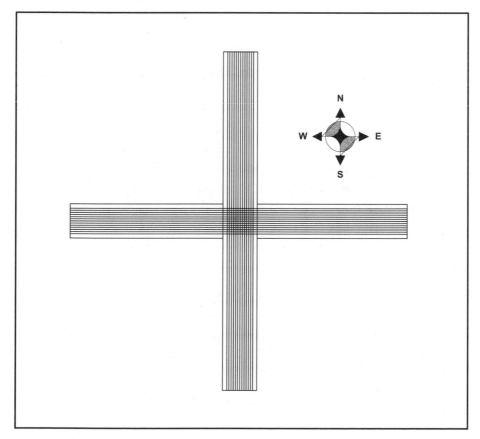

Figure 4-6.
Loop orientation.

Results to Expect

The original article claimed detection to distances of 500 miles, but I doubt that figure is practical. I would more likely guess tens of miles, or 100 miles, but further experimentation is needed to confirm the longer distance claim. *Figure 4-7* shows typical patterns. These patterns are representations of what is observed, but are a bit more distinct and less ragged than actual scope photos. (I am no longer able to photograph scope screens...until I find a new handheld *Polaroid* scope camera.) I was able to create the patterns in *Figures 4-7A* and *4-7B*, but those of *Figures 4-7C* and *4-7D* are cribbed from the Leary article cited earlier.

The pattern in *Figure 4-7A* is what to expect when there is no Z-axis input to receive the vertical sense antenna signal. It shows the line of the storm, but has the directional ambiguity found on loop antennas; in other words, the storm could be in either direction from the loop. *Figure 4-7B* shows the pattern to expect when the sense antenna is used, and the Z-axis intensity is correctly adjusted. This is essentially the same concept as a sense antenna producing a cardioid pattern on a radio direction-finding (RDF) antenna. The actual pattern will be a lot more ragged than shown here. Leary claims that patterns like *Figure 4-7C* are produced with horizontally polarized cloud-to-cloud discharges. (The others were essentially ground wave signals.) The pattern in *Figure 4-7D* represents what might be seen when a ground wave and cloud-to-cloud pattern coexist. I was not able to confirm these patterns, but that might be the particular thunderstorm I observed several years ago when I was active in building loops while researching a forthcoming book (*Joe Carr's Loop Antenna Handbook*, Universal Radio Research, Reynoldsburg, OH).

Electric Field Measurement

Figure 4-8 shows a partial schematic of a sensor used to measure the electrical field at or near ground level produced by thunderstorms. I do not recommend building this project unless you are a professional researcher. And if you do build it, stay a good distance away from it when thunderstorm activity is nearby.

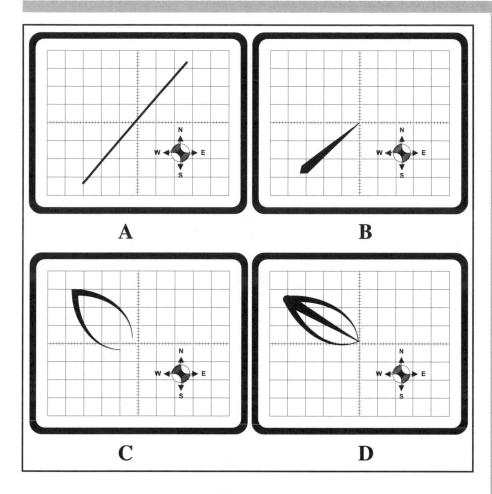

Figure 4-7.
Typical display
results.

Electrical fields are measured in terms of *volts per meter* (V/m), or *kilovolts per meter* (kV/m). The sensor shown in *Figure 4-8* uses a pair of 10-cm copper electrodes on a 4-cm diameter insulated rod, spaced one meter (100 cm) apart. A pair of resistors is used as a voltage divider to reduce the voltage. (Thunderstorm fields of 10 kV/m are easily observed.) The professional sensor I saw used a high-voltage resistor for the upper resistor in the voltage divider (i.e., the type of resistor used inside a high-voltage probe for an oscilloscope or voltmeter). The output of the voltage divider is fed to a high-voltage isolation amplifier, and then to a fiber-optic transmitter. Safety dictates that the sensor be far away from humans and the instruments used to record the burst. Fiber-optic cable is used to prevent the electromagnetic pulse from inducing a high-voltage spike into the wiring.

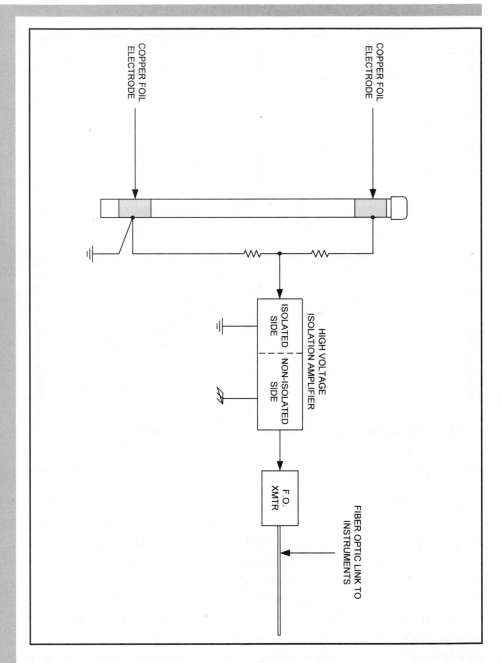

Figure 4-8.
Electrical field
detector.

I saw a primitive variant on this theme at a friend's house in Texas in the
early 1980s. My friend was an antenna guru and something of a technical
mentor for me. He owned a 43-acre tract of farmland near Austin, mostly
for the purpose of erecting antennas at will (but partly to get away from

other people—he was semi-hermit). He had a 1,400-foot long wire antenna mounted on telephone poles, trees and whatever else could be commandeered. At the receiver end of the antenna he had a small box (*Figure 4-9*) with a series-connected stack of ten 1-megohm carbon resistors. The box, and the bottom end of the resistor chain, were grounded.

When I first saw the rig I noted the little light on the box blinking erratically, so asked what it was. He explained to me that he had to do occasional receiver repairs before he realized that so many burned-out RF front-ends could not be due to faulty design, and figured it was atmospheric electric fields "charging" the 1,400-foot wire. He was right. A high-voltage static charge could build up even when the storms were many miles away. He designed this rig after something he'd seen in his years as an electronics technician in the Navy.

Not shown in *Figure 4-9* is a second neon glow lamp connected directly across the antenna terminals. It would protect against spikes, while the resistor stack would "bleed off" static charges. Also, there was a lightning arrester in the antenna downlead.

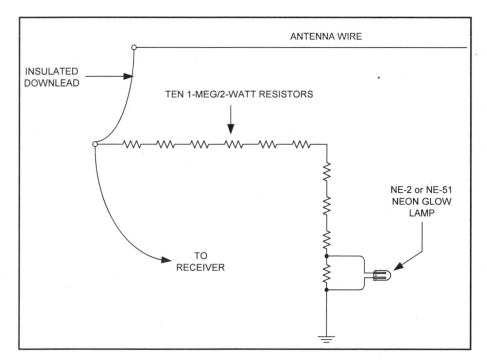

Figure 4-9.
Long-wire antenna
with static discharger.

Chapter 5
ULF/ELF/VLF ELECTRONIC CIRCUITS

Chapter 5
ULF/ELF/VLF Electronic Circuits

A large number of RadioScience Observers work in the frequencies below 100 kHz. The U.S. time and frequency station, WWVB at Fort Collins, CO, operates on 60 kHz. A number of U.S. Navy radio stations operate between 20 and 30 kHz, where they communicate with submarines. A number of radio navigators operate between 10 kHz and 20 kHz. There are also natural radio signals found in these ranges. "Whistlers" and "spherics" occur in the 1-to-10 kHz range. These signals are believed to be produced by lightning strokes at great distances from the receiving site.

A phenomenon called *Shumann resonances* occurs around 8 Hz (not kHz). These are resonances caused by the fact that the space between the earth's surface and the ionosphere acts like a giant waveguide at low frequencies. This accounts for the dramatic rise in VLF signal level (17-30 kHz range) when a solar flare causes the ionosphere to ionize much more heavily down to the D-layer.

A number of these phenomenon were described in *RadioScience Observing Volume 1*, so will not be repeated here. What we will discuss in this chapter are some of the circuits that can be used for making receiver and instrumentation circuits for use in the range 0.1 Hz to, say, 100 kHz.

Operational Amplifier Gain Block

The operational amplifier is a nearly universal gain block that can be used for a variety of low-frequency circuits. Indeed, at one time it was *limited* to low-frequency circuits (some models under 8 kHz). Today, however, one can buy operational amplifiers with gain-bandwidth products—i.e., the frequency at which gain drops to unity—of many hundreds of megahertz (MHz). A number of devices can be used for VLF and below. I have successfully used the 5532 and 5534 devices, as well as the CA-3140 and CA-3160 to frequencies around 100 kHz. (Note: CA-3240 has two CA-3140 devices in the same package, while the CA-3260 has two CA-3160 devices.)

The circuit in *Figure 5-1* is a universal gain block based on any of the operational amplifiers described above. For the single devices above (and all other "industry standard" op-amps), which come in an 8-pin mini-DIP IC package, the pinouts are:

Pin	Function
1	Offset/frequency compensation
2	Inverting input
3	Noninverting input
4	V- DC power
5	Offset/frequency compensation
6	Output
7	V+ DC power
8	Offset/frequency compensation

The functions of pins 1, 5 and 8 vary from one model to another.

The two inputs are inverting (-IN) and noninverting (+IN). The -IN input produces an output signal that is inverted, i.e., phaseshifted 180 degrees, from the input signal. The +IN input produces an output signal that is in-phase (i.e., 0 degrees or 360 degrees) with the input signal. Taken together, the two inputs are a *differential pair*. If signal voltage V1 is applied to the -IN input, and V2 is applied to the +IN input, then the output is proportional to the difference; i.e., V2-V1. Such circuits are said to be *differential amplifiers*.

If a dual operational amplifier (e.g., CA-3240) is used, then the pinouts for its 8-pin mini-DIP package are:

Pin	Function
1	OUT1
2	-IN1
3	+IN1
4	V- DC power
5	+IN2
6	-IN2
7	OUT2
8	V+ DC power

The value of the DC power supply voltages are usually ±6 to ±15 volts, with ±9 to ±12 volts being the most commonly used.

The stages used in *Figure 5-1* are single-ended, i.e., not differential. The signals are applied to either -IN or +IN, but not both. For example, amplifier A1 is the input stage. The signal is applied to the noninverting input, and the amplifier is called a *noninverting follower*. But in A2 and A3 the signal is applied to the -IN input, and the +IN is grounded, the *inverting follower* configuration. Let's look at these two types of stage.

The noninverting follower is at A1. The signal is applied to the non-inverting (+IN) input through a capacitor (C1). A 10 megohm resistor is connected between the +IN terminal and ground. This resistor should be as high as possible, but in any event at least ten times (some prefer 100X) the source impedance. The mid-band gain of the stage is given by:

$$A1 = \frac{R3}{R2} + 1 \qquad (5\text{-}1)$$

Where:

A1 is the gain of stage A1

R2 and R3 are the resistances of R1 and R2.

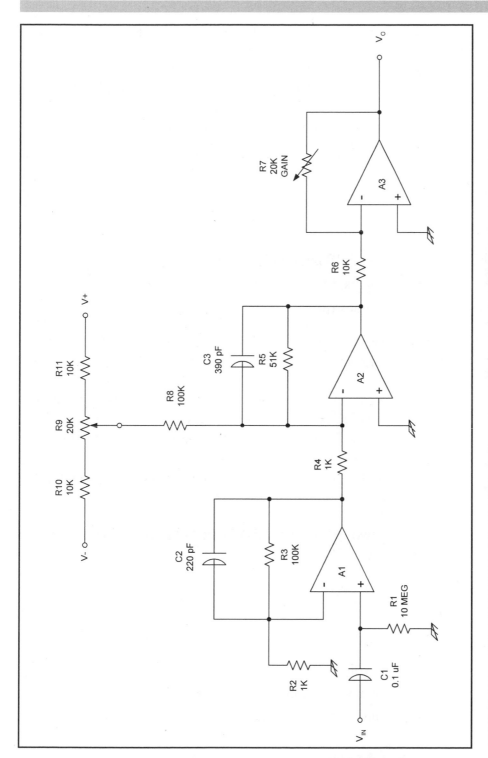

Figure. 5-1.
Low-frequency
amplifier circuit.

With the values shown, R2 = 1K and R3 = 100K, and the gain is (100K/1K) + 1 = 101.

There are two methods used to frequency tailor this circuit. Placing an optional capacitor (C2) across feedback resistor R3 forms a low-pass filter. Similarly, the combined action of C1 and R1 form a high-pass filter. In either case, the -3 dB point on the frequency response is found by:

$$F_{HZ} = \frac{1}{2\pi R C} \tag{5-2}$$

Where:

F_{Hz} is the -3 dB point in Hertz (Hz)

R is in ohms (Ω)

C is in farads (μF)

For the high-pass filter at the input side of A1, Equation 5-2 evaluates to 0.159 Hz, and the low-pass filter the frequency is 7240 Hz.

The use of the capacitor in the input is not strictly necessary, but does serve to block any DC components that might be on the incoming signal. However, it is usually advisable to keep the input resistor unless the signal source is always connected. If the input terminal is allowed to "float" (i.e., no resistor and no signal source), then the op-amp may saturate to the maximum DC output level.

The following two stages (A2 and A3) are both inverting followers. In either case, the gain is:

$$A = -\frac{R_F}{R_{IN}} \tag{5-3}$$

Where:

A is the voltage gain

R_F is the resistor from output to -IN (R5 or R7)

R_{IN} is the resistor from the signal source to -IN (R4 or R6)

Amplifier A2 has a capacitor shunting the feedback resistor, so also operates as a low-pass filter. Equation 5-2 is used to determine the -3 dB point on the frequency response. In this case, $F_{-3dB} = 8,000$ Hz.

Stage A3 is a master gain control for this amplifier chain. It uses a 20 kohm potentiometer for the feedback resistor, and its value can be set from 0 to 20,000 ohms. With a 10 kohm input resistor, therefore, the gain will vary from 0 to -2. The potentiometer can be either a trim pot or a panel-mounted pot, depending on what the requirement is.

DC Balance/Offset Null. Note the other potentiometer in the circuit, R9, and the associated resistors. These components form the *DC offset null* control. If a DC component exists on the signal, or if the DC offsets of the op-amps used in the circuit are large, then the offsets will be amplified as any other signal would. This results in a DC component in the output signal, and if the gain is high even small DC offsets are capable of pegging the output signal to the maximum DC value. This control eliminates the problem.

To adjust R9, set both R7 and R9 to approximately mid-scale, and ground the V_{IN} terminal. Measure the DC output level at V_{O}. It should be zero. Adjust R9 until the level at V_{O} is zero. Next, to ensure the adjustment is proper, run R7 through its entire range while looking at the output voltage. It should not change. Some people prefer to test the circuit by ungrounding V_{IN} and applying a sinewave signal that is well within the frequency and amplitude range of the circuit. Display the output on an oscilloscope set for DC coupling. Adjust R7 from 0 to its maximum value, while looking for baseline shift of the signal. There should be none (the circuit is in "DC balance").

Gain. The overall gain is the product of the three individual stage gains, i.e.:

$$A = A1 \times A2 \times A3 \qquad (5\text{-}4)$$

or,

$$A = \left[\frac{R3}{R2} + 1\right] \times \left[\frac{-R5}{R4}\right] \times \left[\frac{-R7}{R6}\right] \qquad (5\text{-}5)$$

Given the values shown in *Figure 5-1*, the overall gain is:

$$A = 101 \times 51 \times 2 = 10,302 \qquad (5\text{-}6)$$

This gain translates to a gain in decibels of:

$$A_{dB} = 20 \, LOG \, (10,302) = 80.3 \, dB \qquad (5\text{-}7)$$

So how much gain is needed? That depends entirely on the application and the design of your instrumentation system. For example, suppose we want a receiver that will amplify a signal of 200 mmV to a level of 1 volt. The gain required is their ratio; i.e., $1 \, V/2 \times 10^{-4} \, V = 5,000$. The sensitivity required of the receiver, and the maximum desired output voltage level, are the defining characteristics.

DC Power Distribution

Figure 5-2 shows the DC power distribution for any instrumentation system, including the amplifier of *Figure 5-1*. This is a generic circuit, and applies especially to operational amplifier and other circuits that have both V- and V+ DC power supplies.

There are two issues to deal with when designing the DC system. First, you must keep the op-amp stable, and that requires decoupling capacitors right at the DC power terminals of the device. In stage A1 these capacitors are C3 and C4. Typical values are 0.1 µF for the upper end of the range, to 2.2 µF in the ULF region. The other issue is decoupling feedback between the stages, and bypassing noise that comes in on the DC power lines. These functions are partially filled by the decoupling capacitors at the IC terminals. But it is also a good practice to place high-value (220 µF to 1,000 µF) capacitors at each DC power line. The usual practice is to place these capacitors at the point where the DC power is connected to the wiring board. NOTE THE POLARITY. The large-value capacitors (and the others >1 µF) are usually polarized. It is ESSENTIAL that the DC polarity be observed, otherwise the capacitor may explode.

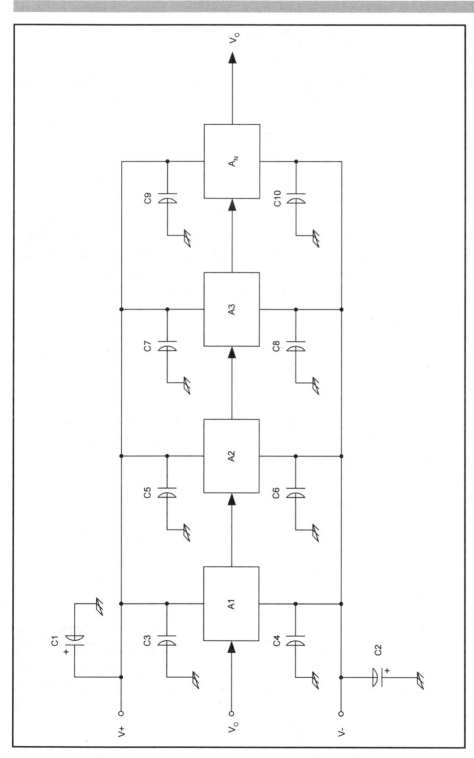

Figure 5-2.
DC power
distribution.

Differential Amplifiers

A differential amplifier produces an output that is proportional to the gain and the difference between the two input signals. If V1 and V2 are the signals applied to the -IN and +IN input, and A_v is the voltage gain, for example, then the output voltage V_o is:

$$V_O = A_v(V2 - V1) \qquad (5\text{-}8)$$

Figure 5-3 shows a pair of differential amplifiers. There are other circuits, but these two examples are capable of receiving signals from high-impedance sources where other circuits are not. The circuit in *Figure 5-3A* is a classic *instrumentation amplifier*. The gain of this circuit is:

$$A_v = \left[\frac{2\,R2}{R1} + 1 \right] \times \left[\frac{R6}{R4} \right] \qquad (5\text{-}9)$$

Where:

A_v is the voltage gain

R1, R2, R4 and R6 are resistance (use the same units for all four).

Equation 5-9 assumes the following:

R2 = R3

R4 = R5

R6 = R7

Resistor R1 is sometimes used as a master gain control, but care must be taken to keep it from dropping to (or even close to) zero. The gain will try to go to infinite, which means the amplifier will saturate immediately. It is common practice to use a fixed resistor in series with a potentiometer for R1 when it is used as a gain control. Set the fixed resistor such that it produces the maximum permissible gain.

Figure 5-3B shows a commercial integrated circuit based on a circuit such as that in *Figure 5-3A*. This IC used a gain setting resistor (R_G) to determine the voltage gain.

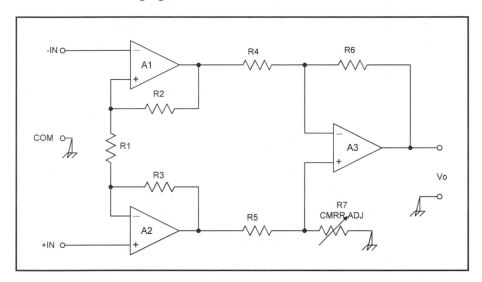

Figure 5-3A.
Op-amp type
instrumentation
differential amplifier.

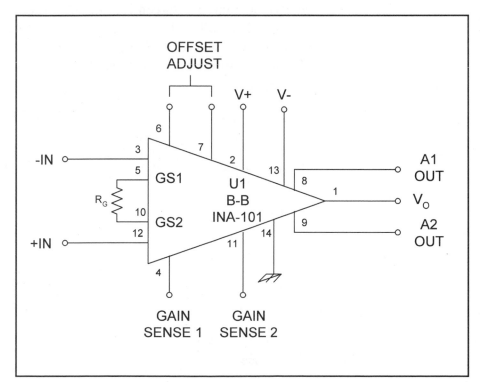

Figure 5-3B.
INA-101 ICIA
instrumentation
differential amplifier.

60 Hz Notch Filter

One of the things that plagues ULF-VLF receivers is 60 Hz interference. Chapter 15 deals with several methods for dealing with 60 Hz interference. But a passive notch filter is shown in *Figure 5-4*. This circuit places a series resonant L-C tank circuit shunting across the signal line. The resonant frequency is found from:

$$F = \frac{10^6}{2\pi\sqrt{L1\ C1}} \qquad (5\text{-}10)$$

Where:

F is the notch frequency in Hertz (Hz)

C1 is the capacitance in microfarads (µF)

L1 is the inductance in Henrys (H)

A series resonant circuit has a low impedance at its resonant frequency, so there will be a "voltage divider" action between R1 and the circuit, but only at the resonant frequency. For a 60 Hz filter, use 0.7 µF and 10 H for the components. Overseas readers, where 50 Hz AC power is used, should use 1.01 µF (i.e., a selected 1 µF unit, or two capacitors in series or parallel that yield 1.01 µF together).

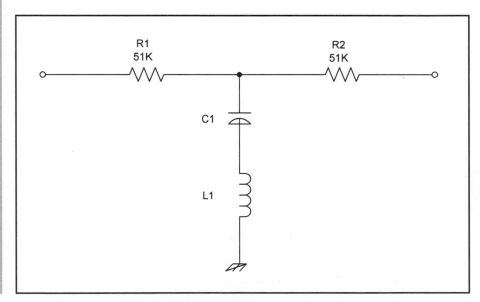

Figure 5-4.
Low-frequency ELF notch filter for 50 or 60 Hz.

Chapter 6
MAGNETOMETER SENSORS

Chapter 6
Magnetometer Sensors

Magnetic fields are of significant interest to scientists and engineers, so it is not surprising that there are a number of magnetic sensors available. Amateur radio operators and shortwave listeners who study propagation often use magnetometers to measure the earth's magnetic field as a function of time. The usual kit used by most amateur observers is a kluge called the "jam jar magnetometer." It works, but its elegance leaves something to be desired (which is a charitable way of putting it!). In this article we will look at a version of the flux gate magnetic sensor that is available to amateurs, plus some applications such as magnetometers and gradiometers. Some of these are useful to the propagation students, while others are used by detectorists, amateur scientists and RadioScience Observers.

Magnetometers are used in a variety of applications in science and engineering. One high-tech magnetometer is used by naval aircraft to locate submarines. Metal detectorists and archeologists use magnetometers to locate buried treasure, marine archeologists and treasure hunters use the devices to locate sunken wrecks and sunken treasure. In industry, magnetometers and related sensor circuits are used to detect anything made of ferrous metal. One such application is counting or detecting the presence of parts on an assembly line.

Gradiometers are differential magnetometers; i.e., a system of two balancing magnetometers, usually in the vertical plane, that produce equal outputs when there are no magnetic anomalies in the vicinity of either sensor.

Flux-Gate Sensors

Flux-gate magnetic sensors are basically overdriven magnetic-core transformers in which the transducible event is the saturation of the magnetic material. These devices can be made very small and compact, yet provide reasonable accuracy.

Figure 6-1 shows the most simple form of flux-gate magnetometer sensor. It consists of a nickel-iron rod used as a magnetic core, wound with two coils. One coil is used as the excitation coil, while the other is used as the output or sensing coil. The excitation coil is driven with a square wave (*Figure 6-2*) with an amplitude that is high enough to saturate the core. The output current signal will increase in a linear manner so long as the core is not saturated. But when the saturation point is reached, the inductance of the coil drops and the current rises to a level limited only by the coil's other circuit resistances.

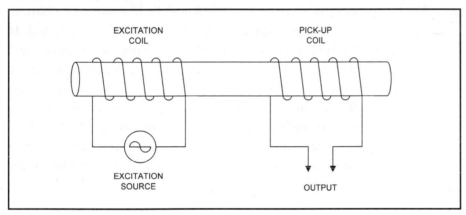

Figure 6-1.
Simple form of flux-gate magnetometer.

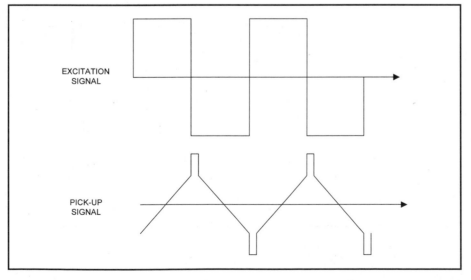

Figure 6-2.
Excitation and output currents.

If the sensor of *Figure 6-1* were in a magnetically pure environment, then the magnetic field produced by the excitation coil would be the end of the story. But there are magnetic fields all around us, and these either add to or subtract from the magnetic field in the core of the flux-gate sensor. Magnetic field lines along the axis of the core have the most effect on the total magnetic field inside the core. As a result of the external magnetic fields, the saturation condition occurs either earlier or later than with only the excitation field in operation. Whether the saturation occurs early or later depends on whether the external field opposes or reinforces the excitation field.

The *variation in entering saturation state* is the transducible event on which this type of sensor is based. Unfortunately, it's also a bit difficult to recover this information. A better solution is shown in *Figure 6-3*. In this version of the flux-gate magnetometer there are two independent cores, each of which has its own excitation winding. A common pickup winding serves both cores. The excitation coils are wound in the series-opposing manner such that the induction currents in the cores precisely cancel each other if the external field is zero. The external field causes pulses to arise in the pickup coil that can be integrated in a low-pass filter to produce a slowly varying DC signal that is proportional to the applied external magnetic field.

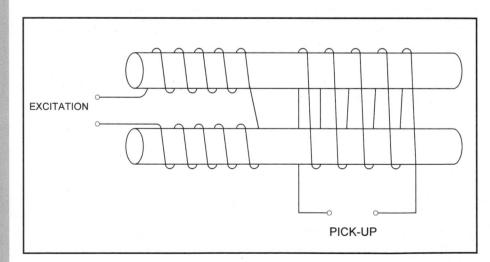

EXCITATION

PICK-UP

Figure 6-3.
Dual core magnetometer is an improvement on the basic design.

Toroidal Core Flux Gate Sensor

The straight core flux-gate sensors suffer from two main problems. First, the desired signal is small compared with the signal on which it rides, so is difficult to discriminate properly. Second, there must be a very good match between the cores and the excitation winding segments on each winding. While these can be overcome, it becomes expensive and thus suffers in popularity.

A better solution is to use a toroidal or "doughnut" shaped magnetic core (*Figure 6-4*). This type of core relieves the problem of picking off small signals in the presence of large offset components. It also reduces the drive levels required from the excitation source.

In the toroidal core flux-gate sensor we can get away with using a single excitation coil wound over the entire circumference of the toroidal core (*Figure 6-4*). The pickup coil is wound over the outside diameter of the core, rather than around the ring as is the excitation coil.

Another advantage of the toroidal core form of magnetometer sensor is that a pair of orthogonal (right angle) pickup cores can be installed that will allow null measurements to be made.

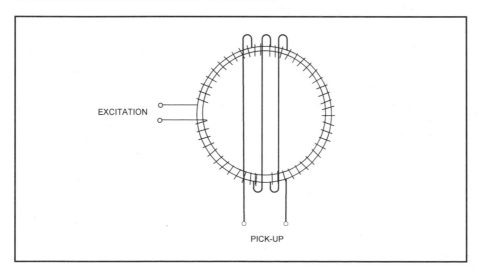

Figure 6-4.
Toroid-core flux gate
magnetometer.

Figure 6-5 shows the orientation of the toroid core flux gate sensor as a function of sensitivity. The maximum sensitivity occurs when the magnetic H-field is orthogonal to the pickup coil, while minimum sensitivity occurs when the pickup coil and H-field are aligned with each other. As you can see, when there are two pickup coils at right angles to each other, then one will be most sensitive as the other goes through null condition.

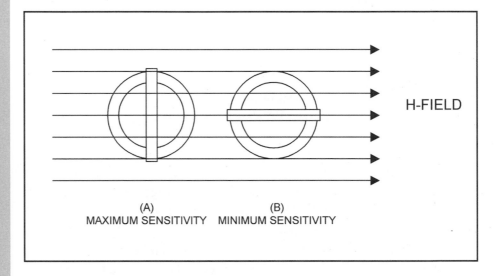

(A)
MAXIMUM SENSITIVITY

(B)
MINIMUM SENSITIVITY

H-FIELD

Figure 6-5.
Orientation effects
on sensitivity.

A Practical Flux-Gate Sensor

A compact and reasonably low cost line of flux-gate sensors, designated FGM-x, is made by Speake & Co. Ltd., and distributed in the United States by *Fat Quarters Software*. The FGM-3 (*Figure 6-6*) device is the one that I evaluated when preparing this article. It is 62 mm long by 16 mm diameter (2.44" x 0.63"). These devices convert the magnetic field strength to a signal with a proportional frequency. One of the things I found fascinating about the FGM-3 is that a set of only three leads provides operation:

Speake & Co. Ltd.
(Elvicta Estate,
Crickhowell, Powys,
Wales, UK)

*Fat Quarters
Software* (24774
Shoshonee Drive,
Murrieta, CA 92562;
909-698-7950 voice;
909-698-7913 FAX)

Red	+5 VDC (Power)
Black	0 Volts (Ground)
White	Output Signal (Square Frequency)

The magnetic detection rating of the device is ±0.5 Oersted (±50 µTesla). This range covers the earth's magnetic field, making it possible to use the sensor in Earth field magnetometers. Using two or three sensors in conjunction with each other provides functions such as compass orientation, three-dimensional orientation measurement systems, and three-dimensional gimballed devices, such as virtual reality helmet display devices. It can also be used to provide magnetometry (including earth field magnetometry), ferrous metal detectors, underwater shipwreck finders, and in factories as conveyer belt sensors or counters. There are a host of other applications where a small change in magnetic field is the important transduction event.

The FGM-3 has the three leads discussed above, and is of the 62 x 16 mm size. The FGM-1 device is smaller than the FGM-3, being 30 mm (1.18") long by 8 mm (0.315") diameter. It has a small connector on one end consisting of four pins: 1) feedback; 2) signal output; 3) ground; and 4) +5 VDC power. The signal, output and ground terminals are essentially the same as on the FGM-3, but the feedback pin provides some extra flexibility. The feedback pin leads to an internal coil that is wound over the flux-gate sensor. It is used to alter the zero field output frequency, or to improve linearity of the sensor over the entire range of the sensor.

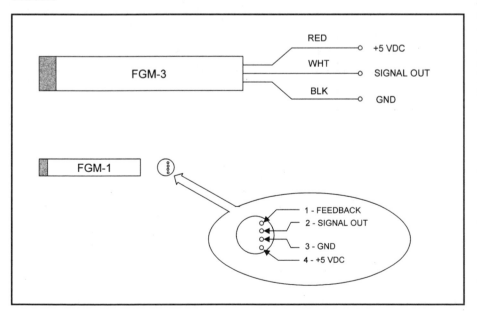

Figure 6-6.
Speake & Co. Ltd.
FGM-1 and FGM-3
sensors.

The FGM-series device output is a +5 volt (TTL compatible) pulse whose period is directly proportional to the applied magnetic field strength. This relationship makes the frequency of the output signal directly proportional to the magnetic field strength. The period varies typically from 8.5 µS to 25 µS (see calibration curve in *Figure 6-7*), or a frequency of about 120 kHz to 50 kHz. For the FGM-3 the linearity is about 5.5 percent over the ±0.5 Oersted range.

The FGM-1, FGM-2 and FGM-3h sensors are related to the FGM-3. The FGM-1 is the smaller version of the FGM-3, with a range of ±0.7 Oersted (± 70 µTesla). The FGM-2 is an orthogonal sensor that has two FGM-1 devices on a circular platform at right angles to one another. This orthogonal arrangement permits easier implementation of orientation measurement, compass and other applications. The FGM-3h is the same size and shape as the FGM-3, but is about 2.5 times more sensitive. The output frequency changes approximately 2 to 3 Hz per gamma of field change, with a dynamic range of ±0.15 Oersted (about one-third the Earth's magnetic field strength).

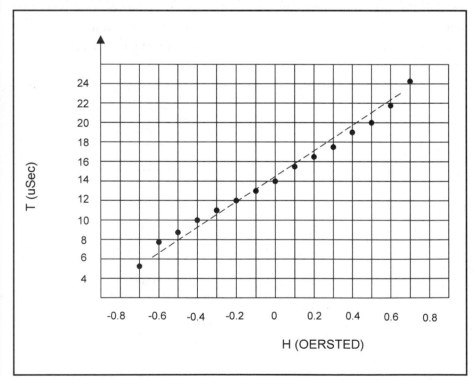

Figure 6-7.
Calibration points
on a typical FGM-3.

The response pattern of the FGM-x series sensors is shown in *Figure 6-8*. It is a "figure-8" pattern that has major lobes (maxima) along the axis of the sensor, and nulls (minima) at right angles to the sensor axis. This pattern suggests that for any given situation there is a preferred direction for sensor alignment. The long axis of the sensor should be pointed towards the target source. When calibrating or aligning sensor circuits, it is common practice to align the sensor along the east-west direction in order to minimize the effects of the Earth's magnetic field.

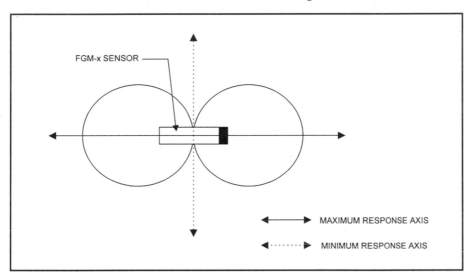

Figure 6-8. Sensitivity pattern for the FMG-x devices.

Powering the Sensors

The FGM-x series of flux-gate magnetic sensors operates from +5 volts DC, so is compatible with a wide variety of analog and digital support circuitry. As is usual for any sensor, you will want to use only a regulated DC power supply for the FGM-x devices. In fact, the manufacturer recommends that double-regulation (*Figure 6-9A*) be used. Ripple in the DC power supply can cause output frequency anomalies, and those should be avoided. In the circuit of *Figure 6-9A*, an unregulated +12 to +15 VDC input potential is applied first to a 9-volt 78L09 or 78M09 three-terminal IC voltage regulator (U1). This produces a +9-volt regulated potential that is then applied to the input of the 78L05 or 78M05 device (U2). The second regulator reduces the +9 volts from U1 to the +5 volts needed for the FGM-x sensors.

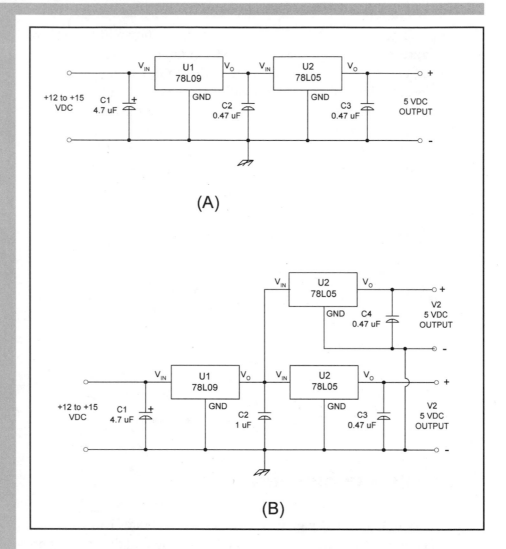

Figure 6-9.
A) Double-regulated
DC power supply
for FGM-x sensors;
B) supply to use
when other circuits
must be powered.

When other digital devices are being powered from the same DC power supply it is prudent to provide a separate DC source for the FMG-x sensors. In *Figure 6-9B* we see the type of circuit that would accomplish this task. There are two separate +5 VDC outputs, labeled V1 and V2. Both are derived from 78L05 devices that are powered from a single 78L09. Care must be taken to not exceed the maximum current limits of U1, especially if the same size IC voltage regulators are used for all three (U1, U2, and U3). One of the +5 VDC sources, either V1 or V2, can be used for powering the FGM-x device, while the other powers the rest of the circuitry.

Calibration of the Sensors

The FGM-x devices are not precision instruments out of the box, but can be calibrated to a very good level of precision. The calibration chore requires you to generate a precise magnetic field in which the sensor can be placed. One way to generate well-controlled and easily measured magnetic fields is to build a coil and pass a DC current through it. If the sensor is placed at the center of the coil (inside), then the magnetic field can be determined from the coil geometry, the number or turns of wire and the current through the coil. There are basically two forms of calibrating coil found in the various magnetic sensor manuals: *solenoid-wound* and the *Helmholtz pair*.

Figure 6-10 shows the solenoid coil. A solenoid is a coil that is wound on a cylindrical form in which the length of the coil (L) is greater than or equal to its diameter. This type of coil is familiar to radio fans because it is used in many L-C tuning circuits. The magnetic field (H) in Oersteds is found from:

$$H = \left(\frac{4\pi \, N \, I \, L}{10 \; SQR(\, L^2 + D^2 \,)} \right) \tag{6-1}$$

Where:

H is the magnetic field in Oersteds

N is the number turns-per-centimeter (t/cm) in the winding

I is the winding current in amperes

D is the mean diameter of the winding in centimeters (cm)

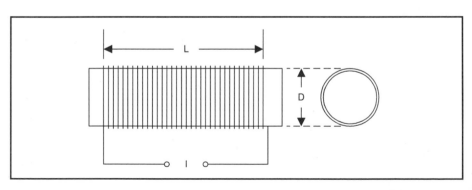

Figure 6-10.
Solenoid-wound
calibration coil.

The winding is usually made with either #24 or #26 enameled or <u>Formvar</u>® covered copper wire. The length of the solenoid coil should be at least twice as long as the sensor being calibrated, and the sensor should be placed as close as possible to the center of the long axis of the coil.

The Helmholtz coil is shown in *Figure 6-11*. It consists of two identical coils (L1 and L2) mounted on a form with a radius R, and a diameter 2R. The coils are spaced one radius (1R) apart. The equations for this type of calibration assembly are:

$$H = \frac{0.8991\ N\ I}{R} \qquad (6\text{-}2)$$

and,

$$B = \frac{9.1 \times 10^3\ N\ I}{R} \qquad (6\text{-}3)$$

In the practical case, one usually knows the dimensions of the coil, and needs to calculate the amount of current required to create a specified magnetic field. We can get this for the Helmholtz pair by rearranging *Equation 6-2*:

$$I = \frac{R\ H}{0.8991\ N} \qquad (6\text{-}4)$$

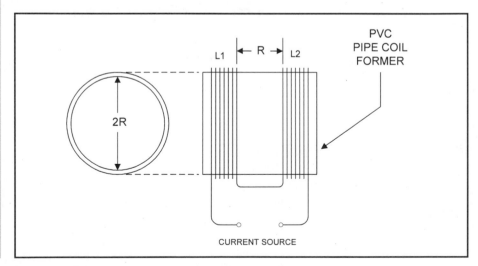

Figure 6-11.
Helmholtz pair
calibration coil.

The coils are a little difficult to wind, especially those of large diameter (e.g., 4-inch). One source recommends using double-sided tape (the double-sticky stuff) wrapped around the form where the coils are to be located. As the wires are laid down on the form they will stick to the tape, and not dither around.

The above equations, plus a lot of magnetic theory and calibration suggestions, plus information on other sensors, are found in Janicke (1994).

Figure 6-12 shows a type of assembly that can be used for either the solenoid or Helmholtz coil. I first saw this type of assembly in a college freshman physics laboratory about 25 years ago. It consists of a PVC pipe section used as the coil former. Endcaps on the coil former also serve as mountings. The mounts at either end consist of smaller segments of PVC pipe and nylon (nonmagnetic) hardware fasteners. Another segment of PVC pipe, of much smaller diameter than the coil former, is passed through the former from one endcap to the other, such that its

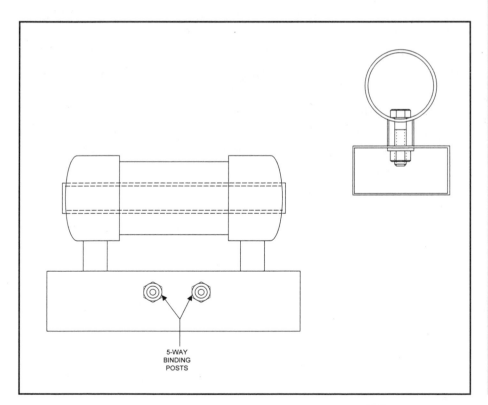

5-WAY
BINDING
POSTS

Figure 6-12.
Coil-mounting
assembly.

ends protrude to the outside. This pipe forms a channel into which the sensor can be placed. The base is a plastic or wooden box (again, non-magnetic materials). One thing nice about this type of assembly is that the sensor is always in approximately the same position in the coil, close to the center of the field.

Analog Interface to FGM-3

Figure 6-13 shows a method for providing an analog interface to the FGM-3 and its relatives. The output of the sensor is a 40 to 125 kHz frequency that is proportional to the applied magnetic field. As a result, we can use a frequency-to-voltage (F/V) converter, such as the LM-2917, to render the signal into a proportional DC voltage. That voltage, in turn, can be used to drive an analog or digital voltmeter or milliamme-ter. The LM-2917 is selected because it is widely available at low cost from mail-order parts distributors.

The output circuit consists of a bridge made up of R1, R2 and R3, along with the output of the LM-2917 device. R2/R3 form a resistive voltage divider that produces a potential of 1/2 volt (V+) at one end of the 22K sensitivity control (R4). If the voltage produced by the LM-2917 is the same as the voltage at R4, then the differential voltage is zero and no current flows. But if the LM-2917 voltage is not equal, then a difference exists and current flows in R4/M1. That current is proportional to the applied magnetic field. Meter M1 and R4 can be replaced with a digital voltmeter, if desired.

The DC power supply uses two regulators, one for the FGM-3 and one for the LM-2917. Even better results can be obtained if an intermediate voltage regulator, say a 78L09, is placed between the V+ source and the inputs of U2 and U3. That results in double regulation, and produces better operation.

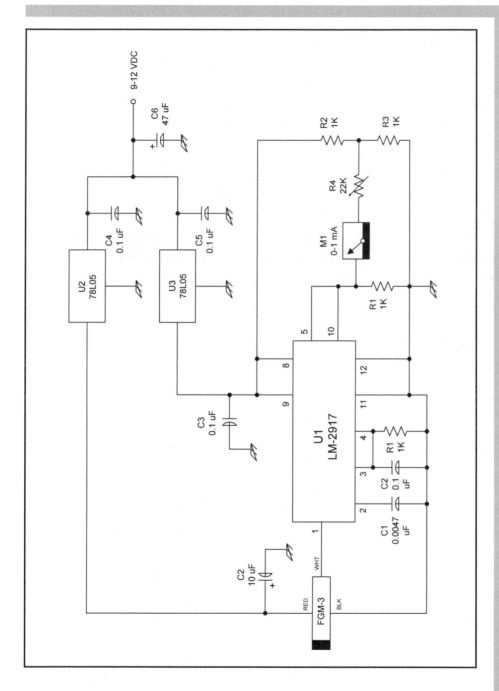

Figure 6-13.
Analog magnet-
ometer using the
LM-2917 F/V
converter.

Digital "Heterodyning"

The method shown in *Figure 6-14* results in a more sensitive measurement over a small range of the sensor's total capability. The circuit makes it possible to measure small fluctuations in a relatively large magnetic field.

A Type-D flip-flop is used in *Figure 6-14* to "mix" the frequency from the FGM-3 with a reference frequency (F_{REF}). The FGM-3 literature calls this process "digital heterodyning" (Dare we call it "digidyning?"), although it is quick to point out that it is really more like the production of alias frequencies by undersampling than true heterodyning.

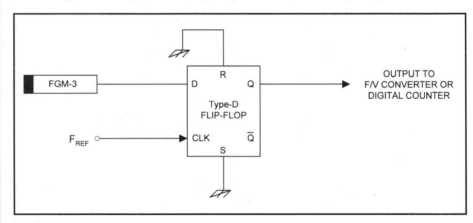

Figure 6-14.
Digital heterodyning
("digidyning") circuit.

Two types of frequency source can be used for F_{REF}. For relatively crude measurements, such as a passing vehicle detector, the CMOS oscillator of *Figure 6-15A* is suitable. This circuit is based on the 4049 hex inverter chip connected in an astable multivibrator configuration. The exact frequency can be adjusted using R2, a 10 kohm, 10-turn potentiometer.

Where a higher degree of stability is needed, for example when making Earth field variation measurements or testing magnetic materials, a more stable frequency source is needed. In that case, use a circuit such as in *Figure 6-15B*. This circuit uses a crystal controlled oscillator feeding a binary divider network. Crystal oscillators can be built, or if you check the catalogs you will find that a large number of frequencies are available

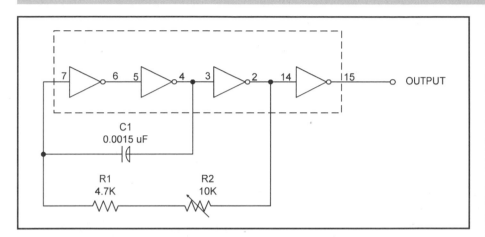

Figure 6-15A.
CMOS reference
oscillator.

in TTL and CMOS compatible formats at low enough costs to make you
wonder why you would want to build your own.

The reference frequency is adjusted to a point about 500 Hz below the
mean sensor frequency. This frequency is measured when the sensor is in
the east-west direction. This arrangement will produce a frequency of
0 to 1,000 Hz over a magnetic field range of ±500 gamma.

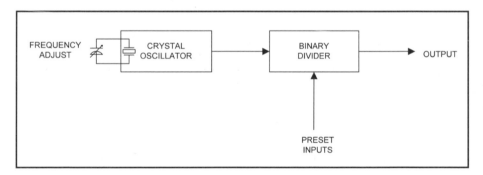

Figure 6-15B.
Crystal reference
source.

A Magnetometer Project & Kit

Figure 6-16 shows the circuit for a simple magnetometer based on the
FGM-3 flux-gate sensor. It can be obtained in kit form from Fat Quarters Software. The connections to the printed-circuit board are shown in
Figure 6-17. This device takes the output frequency of the FGM-3, passes
it through a special interface chip (U1) and then to a digital-to-analog
converter to produce a voltage output.

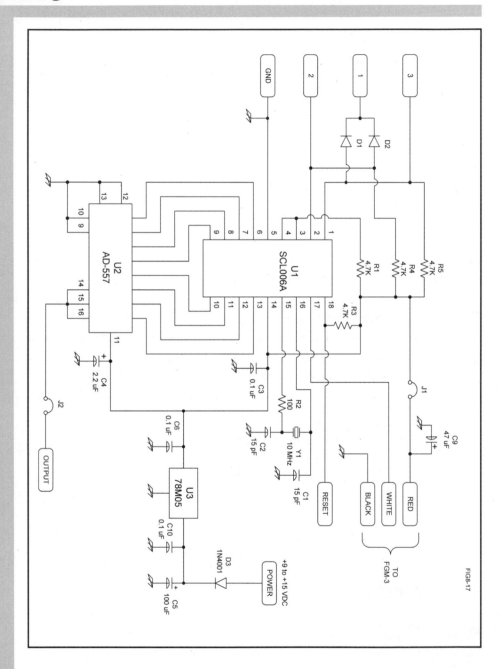

Figure 6-16.
Magnetometer
project circuit.

The sensitivity switch (S1) produces the following sensitivities when the FGM-3 sensor is used:

S1 Position	Sensitivity
4	± 150 gamma
3	± 250 gamma
2	± 550 gamma
1	± 1,000 gamma

These ranges translate to a DC output voltage between 0 and +2.5 volts. If the FGM-3h sensor is used instead of the FGM-3, then divide the sensitivity figures by 2. These figures are approximate. If greater accuracy is needed, then each sensor should be individually calibrated.

The heart of this magnetometer project (*Figures 6-16* and *6-17*), other than the FGM-3 device, is the special interface chip, Speake's SCL006 device. It provides the circuitry needed to perform magnetometry, including Earth field magnetometry. It integrates field fluctuations in one-second intervals, producing very sensitive output variations in response to small field variations. It is of keen interest to people doing radio propagation studies, and who need to monitor for solar flares. It also works as a laboratory magnetometer for various purposes. The SCL006A is housed in an 18-pin DIP IC package.

The D/A converter (U2) is an Analog Devices type AD-557. It replaces an older Ferranti device seen in the Speake literature because that older device is no longer available. Indeed, being a European device it was a bit hard to find in unit quantities required by hobbyists on this side of the Atlantic. The kit from Fat Quarters Software contains all the components needed, plus a printed-circuit board. The FGM-3 device is bought separately.

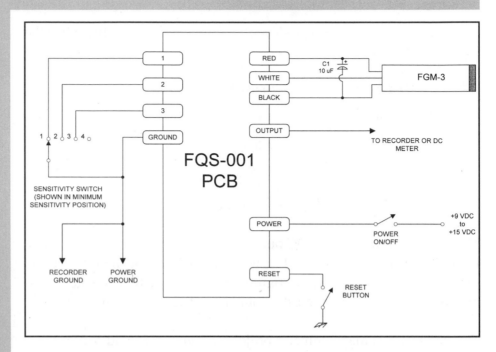

Figure 6-17.
PCB connections for
the magnetometer.

The external connections are shown in *Figure 6-17*. The circuit is designed to be run from 9-volt batteries so that it can be used in the field. A sensitivity switch provides four positions, each with a different overall sensitivity range. The output signal is a DC voltage that can be monitored by a strip-chart or X-Y paper recorder, voltmeter, or fed into a computer using an A/D converter.

If you intend to use a computer to receive the data, then it might be worthwhile to eliminate the D/A converter and feed the digital lines (D0-D7) from the SCL006A directly to an 8-bit parallel port. Not all computers have that type of port, but there are plug-in boards available for PCs, as well as at least one product that makes an 8-bit I/O port out of the parallel printer port.

Sensor Head Mechanical Construction

When evaluating the FGM-3 sensor I built a magnetometer based on *Figures 6-16* and *6-17*, using the kit provided by Fat Quarters Software. The printed circuit, switches and meter were mounted on a small sloping front cabinet. The goal was to build a sensor head that could be rotated to find the magnetic field. (The FGM-x sensors are direction sensitive.) The solution was to place the sensor inside a 0.75-inch (19-mm) PVC plumbing "tee" connector (*Figure 6-18A*). Three endcaps were provided, one for each port on the "tee" connector. The endcap on the downstroke of the "tee" is fitted with a 0.25-inch (6.35-mm) stereo phone plug (see detail in *Figure 6-18B*). When this plug is mated with a phone jack on the top of the project's case, it can be rotated at will.

Mounting the FGM-3 sensor inside the PVC "tee" connector is shown in *Figure 6-18C*. The sensor is mounted horizontally in the crosspiece of the "tee," while the wires are routed to the downstroke section. The sensor is held centered in the cylindrical PVC "tee" with small plugs

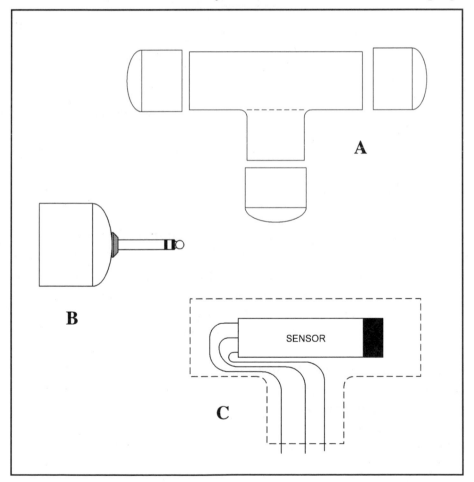

Figure 6-18.
A) PVC tee;
B) endcap with
stereo phone plug;
C) sensor orientation
inside tee.

made of styrofoam or some other material. I used a small hobbyist razor knife to carve the larger-size styrofoam "peanuts" of the sort used for packing fragile items for shipping. The finished sensor assembly is mounted on top of the project's case (*Figure 6-18D*).

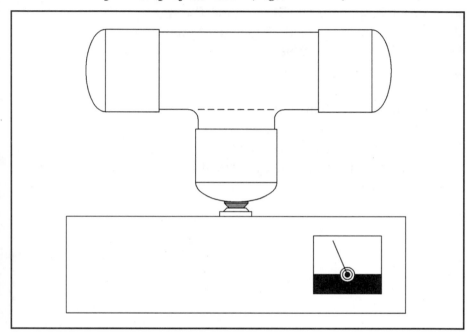

Figure 6-18D. Finished sensor assembly.

Gradiometers

One of the problems with magnetometers is that small fluctuations occur in otherwise very large magnetic fields. And those fluctuations can sometimes be important. A further problem with single-sensor systems is that they are very sensitive to orientation. Even a small amount of rotation can cause unacceptably large, but spurious, output changes. The changes are real, but are not the fluctuations that you are seeking.

A *gradiometer* is a magnetic instrument that uses two identical sensors that are aligned with each other so as to produce a zero output in the presence of a uniform magnetic field. If one of the sensors comes into contact with some sort of small magnetic anomaly, then it will upset the balance between sensors, producing an output. The gradiometer gets its name from the fact that it measures the gradient of the magnetic field over a small distance (typically 30 to 150 cm).

These instruments can be used for finding very small magnetic anomalies, for example, the metallic firing mechanism of plastic land mines buried a few inches below the surface, or a shipwreck buried deep in the ocean silt. Archeologists use gradiometers to find artifacts, and identify sites. Also, people who explore Civil War battlefields, western mining camps, and other sites often use gradiometers to facilitate their work.

Figure 6-19 shows the construction details for a simple gradiometer based on the FGM-3 device. It is built using a length of PVC pipe. One sensor is permanently mounted at one end of the pipe, using any sort of appropriate nonmagnetic packing material. In one experiment, I used the standard 0.5-inch (11-mm) adhesive-backed window sealing tape used in colder areas of the country to keep the howling winds out of the house in wintertime. It worked nicely to hold the permanent sensor in place.

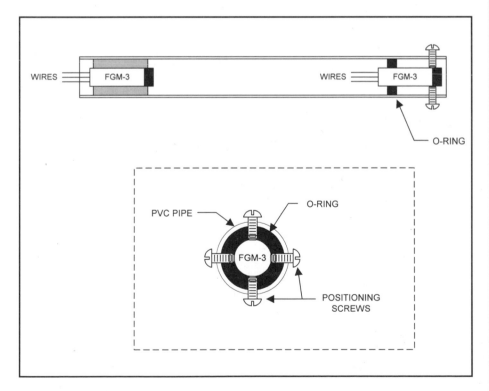

Figure 6-19.
Gradiometer
instrument.

The other sensor is mounted in the opposite end of the tube, using an O-ring that fits snugly into the tube. Four positioning screws made of nonmagnetic materials are used to align the sensor. The position of the sensor is adjusted experimentally. The idea is to position the sensor such that the gradiometer can be rotated freely in space without causing an output variation.

The gradiometer sensor is usually held vertically, such that the end with the wires coming out of the FGM-3 devices is pointed downwards. This allows you to find buried magnetic objects even if they are quite small.

A practical gradiometer can be built using a special interface chip by Speake, the SCL007 device (*Figure 6-20*). It is an 18-pin device that accepts the inputs from the two sensors, and produces an 8-bit digital output. It can receive the signals from the sensors in *Figure 6-19*, and

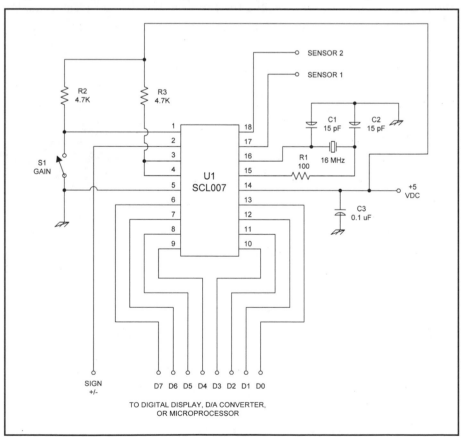

Figure 6-20.
Gradiometer circuit
using the SCL007
device.

produce a digital output proportional to the field gradient. Also, if you want a DC output, then the same sort of D/A converter used in the magnetometer of *Figures 6-16* and *6-17* can also be used for the gradiometer.

The method of digital heterodyning (shown in *Figure 6-21*, and earlier in *Figure 6-14*) can be used to make a very sensitive gradiometer at low cost. The outputs of the two FGM-3 sensors are fed to the D-input and clock (CLK) input of a Type-D flip-flop. The output of the Type-D flip-flop is fed to an F/V converter, such as the LM-2917 device discussed earlier.

Interfacing FGM-x Series Devices via Microcontrollers

Microcontroller chips bring some of the advantages of digital computers in single integrated-circuit or small assembly form. *Figure 6-22* shows a method for interfacing the FGM-x/SCL-00x series of devices to a device such as the Parallax, Inc. *BASIC Stamp*, or the Micromint *PicStic* product. If the calculations cannot be done in the microcontroller, then use the serial output capability to send the 8-bit data to a personal computer.

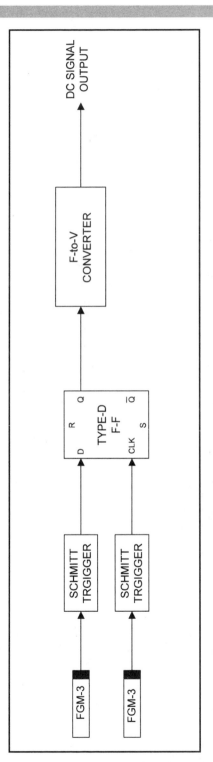

Figure 6-21. Digidyning gradiometer.

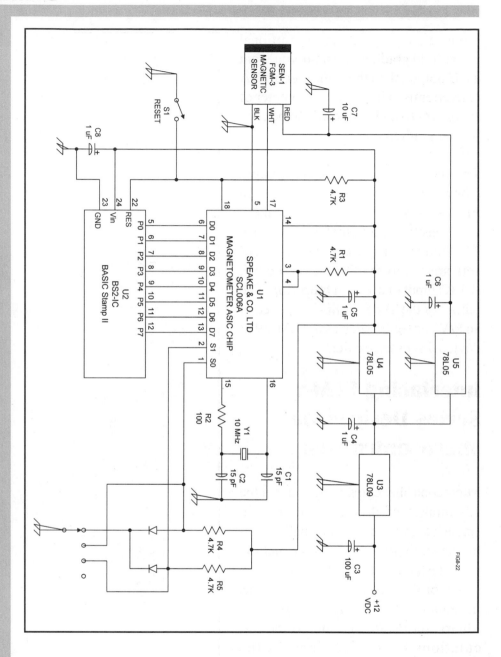

Figure 6-22.
Interfacing FGM-x
series sensors to
microcontrollers.

The Basic Stamp and PicStic can be programmed in a limited version of BASIC. For most instruments I suspect that the program for a magnetometer will be a simple program that inputs data from the binary output of the SCL-006 (or other) interface chip, and then reflects it to the serial input of the computer.

Acknowledgments

I wish to thank the late Mr. Richard Noble of Speake & Co. Ltd. in Wales, UK, and Mr. Erich Kern of Fat Quarters Software in USA, for assistance in preparing this article.

References

Janicke, J.M. (1994). *The Magnetic Measurements Handbook*, Magnetic Research Press, 122 Bellevue Avenue, Butler, NJ, 07405.

Kern, Erich (1996). Fat Quarters Software, 24774 Shoshonee Drive, Murrieta, CA 92562; 909-698-7950 (voice) and 909-698-7913 (FAX). Telephone consultation plus FGM-x literature and drawings.

Noble, Richard (1991), Speake & Co. Ltd (Elvicta Estate, Crickhowell, Powys, Wales, UK. "Fluxgate Magnetometry," *Electronics World and Wireless World*, Sept 1991, pp. 726-732.

Chapter 7
SNIFFING OUT LOCAL SIGNAL
AND NOISE SOURCES

Chapter 7
Sniffing Out Local Signal and Noise Sources

There are numerous local sources of radio signals. Some of these are interesting to research in and of themselves, but for the most part they simply mess up the rest of your RadioScience hobby. The idea is to seek them out and suppress them. But finding some radio frequency (RF) noise problems can be a pain in the anatomy. The stuff seems to radiate from everywhere. The trick is to locate the local source, and from there you can devise strategies to defeat it (see Chapter 15).

Not all sources of electromagnetic interference are caused by intentional RF generators. Although some devices, such as transmitters, induction heaters, and even receiver local oscillators, can leak into the atmosphere, they are not the only problem. Almost any electrical device can cause problems.

When I first moved into my present home my HF receiver was all but useless because of the high "hash" level of noise present. The problem turned out to be a dimmer switch that replaced the ordinary light switch. Those devices use a duty-cycle SCR circuit to lower the power level of incandescent lights, and the truncated waveform they produce is rich in harmonics (well into the HF bands!). I replaced all six dimmers with conventional switches.

Some years ago I served as a volunteer technical representative for the local television interference committee. We had the responsibility for finding the source of interference and recommending solutions. The Federal Communications Commission Field Engineering Office often deferred ham radio and Citizen's Band complaints to the volunteer committee rather than "...making it official" (unless there was a suspicion of illegal radio operations!). During that two-year period I found a large number of non-radio sources of interference to AM BCB, FM BCB, high-fidelity audio and television equipment. Included were: electric motors, a microwave oven, loose tie-wires on AC mains transformers (on the pole outside the home) and electrical space heaters.

Many different forms of appliance were indicted, including (oddly enough) a dishwasher that has an SCR controller inside to turn it on and off, a garbage disposal unit under the complainant's sink, a garage door opener, and the most raucous door bell-chime I've ever heard.

During another period of my life I worked installing both CB and landmobile two-way radios, as well as ordinary automobile radios. One of the main jobs for an installer of mobile electronic gear is to locate and suppress interference sources. And vehicles abound in such sources! The ignition and the charging system are prime culprits, but also causing problems are things like the gas-gauge sending unit, power windows, and almost anything else electrical.

Even if your field of interest is limited to eliminating mobile ignition system noise, the task can be daunting. I've seen cases where an ungrounded hood caused massive noise problems. In another case, noise was induced on the DC power lines that pass through the firewall from the engine compartment to the passenger compartment, where it was reradiated and picked up by the electronic equipment. Also, an ungrounded engine exhaust pipe can reradiate noise.

All of those things are routinely found. But some are not so routine. Once (if you will permit me a nostalgic regression), I was working after school installing CB sets at the dawn of the CB era (late 1950s). The vehicle was a 1956 Ford *Crown Victoria* sedan. I tried everything in the technician's bag of tricks, and the CB kept clicking with ignition noise all across the band. The master technician, a rough-and-ready fellow named Nelson, came down to the garage, determined to "...show that Carr kid how it's done." He inspected my work and could find no fault. He tried a few things himself, and after two hours was still unsuccessful. At that point, weary from lack of success (not to mention a two-hour butt-chewing by Nelson), I leaned my elbow against the chrome roofline of the Crown Vicky. The noise stopped! One of the features that distinguished the '56 Crown Vicky from less costly models was a 9-foot long curved chrome decoration strip around the front of the roofline, continuing on to the two sides of the vehicle. Get the point? Nine-feet is quarter wavelength at the 11-meter (27 MHz) Citizen's Band, so even minute amounts of radiation would find a resonant situation and reradiate right into the antenna! Cleaning and resetting the clips and screws that held the chrome strip fast solved the problem.

Bag of Tricks

The methods of eliminating the noise vary so much from problem to problem that it would take a book to cover them all. The point here is to provide you with some aids to sleuthing the problem; i.e., finding the noise source. Only after the source is located can you effectively devise a strategy to solve the problem. (Filtering, screening, isolation and three sticks of dynamite at 0300 have all been successfully used—although the latter is beyond my experience.)

RF Sleuthing Tools

The correct tool for finding RF sources is anything that will permit you to unambiguously determine where the radiation is coming from. We will take a look at several low-cost possibilities.

If you are looking for a source that is out in the neighborhood somewhere, such as a loose power line or a malfunctioning electrical system in another building, then an ordinary solid-state portable radio may do the trick. Open the radio and locate the loopstick antenna. If you rotate the radio through an arc where the loopstick is first broadside to the arriving signal, and then perpendicular to the signal, you will notice a tremendous reduction in signal. Ferrite loopsticks are extremely sensitive to direction of arrival, with a sharp null occurring off the ends.

By noting the direction in which the null occurs (*Figure 7-1*), you find the line of direction to/from the signal source. It is not unambiguous, however. To find the actual direction from the two possibilities that are

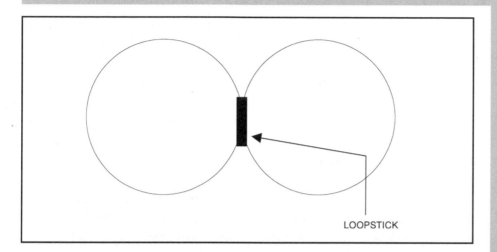

Figure 7-1.
Directional pattern
of a ferrite loopstick
antenna.

180 degrees opposed, move along the line and note in which direction the signal increases. Except in a very few cases where reflections occur, the signal will become stronger as you move towards the source.

The standard medium wave and shortwave loop antenna used by a lot of radio enthusiasts is also useful for finding RF emitting sources. A square loop of 30 to 90 cm to the side is relatively easy to construct, and forms a very directional antenna. Indeed, these antennas are commonly used in radio direction finding. Such loop antennas offer a null when pointed broadside to the signal direction, and a peak when orthogonal to the signal direction.

A portable radio with an S-meter will work wonders in this respect. An ad-hoc S-meter can be formed using the circuit in *Figure 7-2*. This circuit plugs into the earphone jack of the receiver. The received audio is rectified by D1/D2 (a voltage doubler), and then applied to a DC current meter.

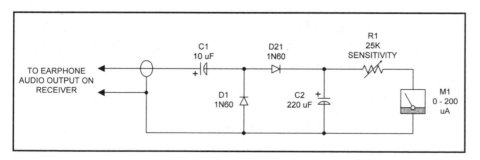

Figure 7-2.
"Audio S-meters"
for use on
earphone-equipped
receivers.

If the noise peaks in the HF shortwave bands, then the radio's loopstick is of little use. (It only works on the MW and LF AM BCBs.) The HF antenna in those receivers is a telescoping whip. A loopstick sensor can be fashioned in the manner of *Figure 7-3*. An 18-cm (7-inch) ferrite rod is wound with about 20 turns of fine enameled wire. This wire is connected to a coaxial cable, that is in turn connected to the external antenna terminals of the receiver (if such exists). If there are no external antenna terminals, then a small coil of several turns wrapped around the telescoping antenna will couple signal to the radio. It's a good idea to keep the antenna at minimum height in order to minimize pickup from sources other than the loopstick. More sensitivity can be had if a resonant loopstick is available, but those also restrict the frequency band.

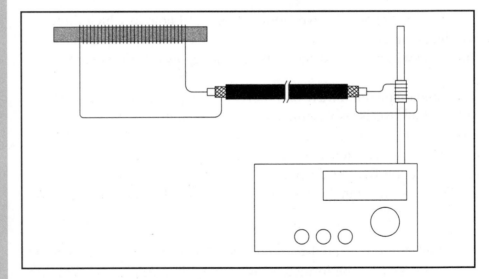

Figure 7-3.
Using a ferrite rod broadband antenna with a telescoping whip antenna on a receiver.

A *Rif Sniffer* that is popular with mobile radio installers is shown in *Figure 7-4*. Although fancy commercial models exist for a price, the basic form of *Figure 7-4* can be made with a length of coaxial cable connected to the receiver antenna terminals. The shield is cut back a distance of 5 to 7 cm, and then removed. The inner conductor, covered with the inner insulator, becomes the sensor for finding RF sources such as ignition noise radiators.

In most cases, some kind of insulating cap is placed over the end of the inner conductor to keep it from contacting voltage sources as it is used

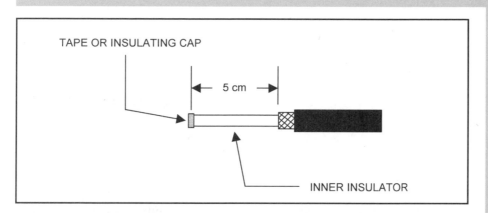

TAPE OR INSULATING CAP

5 cm

INNER INSULATOR

Figure 7-4.
Simple coax sniffer.

to probe for RF. An accidental contact with the 12-volt battery of a vehicle can cause destruction of the input coil on the receiver.

A "gimmick coil" sensor is shown in *Figure 7-5*. The sensor in this case is a solenoid-wound inductor connected to a length of coaxial cable. The coil has a diameter of 2.5 to 5 cm, and a length that is at least its own diameter. The coil consists of two to 10 turns of wire, depending on frequency (the higher the frequency, the fewer the turns required). The other end of the coaxial cable is connected to a receiver. When the coil is brought nearer the noise source, the signal level in the receiver will get higher.

A single-turn gimmick is shown in *Figure 7-6*. This sensor consists of a single loop, approximately 10 cm in diameter, connected to coaxial cable. It can be used well into the low end of the VHF spectrum, as well as at HF. The loop is made of either small-diameter copper tubing, heavy brass wire or rod, or heavy-gauge solid copper wire (#10 or lower). The loop has some directivity, so can be used to ferret out extremely localized sources.

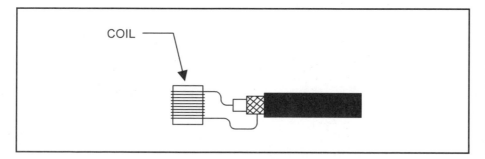

COIL

Figure 7-5.
"Gimmick" coil sniffer.

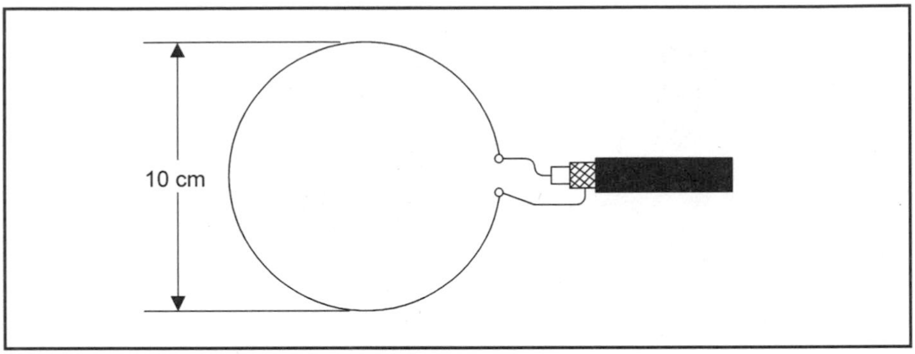

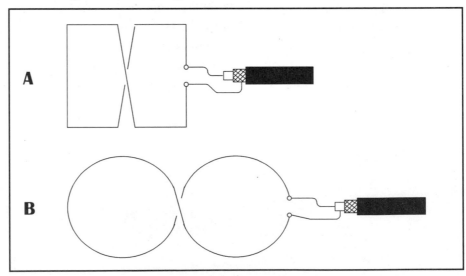

Figure 7-6.
Single-turn loop
sniffer.

A fault with the single-turn loop is that it is somewhat sensitive to magnetic fields, although not so much as some of the other forms. *Figures 7-7A* and *7-7B* show dual-loop sensors that are less sensitive to magnetic field pickup. In *Figure 7-7A* the loops of the sensor are rectangular, and crossed in the center. The feedline (coaxial cable) is connected to a break in one of the two coils.

The version shown in *Figure 7-7B* is circular. One version that I built was made of #8 AWG solid copper wire formed around a *John McCann Irish Oatmeal* tin can. Getting the coils about the right size and reasonably circular is relatively easy given a proper former. The exact size of the coils in *Figures 7-7A* and *7-7B* is not terribly important. The coil should not be too large, however, or it will be less local and may become a bit ambiguous in locating some sources.

A

B

Figure 7-7.
Two forms of anti-
magnetic field
sniffer.

Figure 7-8 shows a method for sensing RF flowing in a conductor. This method is used as the sensor in a lot of ham radio RF power meters and VSWR meters. The conductor is passed through a toroid coil form, essentially acting as a one-turn primary winding of the transformer. The "secondary" winding is made of #22 to #30 wire wound over the toroid. About 6 to 20 turns are used, depending on frequency.

One popular method for constructing the sensor in *Figure 7-8* is to order a toroid core that has an inside diameter a little less than the outside diameter of a rubber grommet. Mount the grommet in the center hole of the toroid core, and then pass the wire through the center hole of the grommet. The two ends of the secondary winding are connected to the inner conductor and shield of the coaxial cable to the receiver.

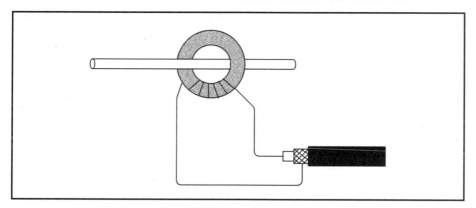

Figure 7-8.
Toroid sensor for RF current carrying wire.

RF Detectors

The RF output of the sensor coils can be routed to a receiver, and for low-level signals may well have to be so treated. For higher power sources, however, an RF detector probe is used. *Figures 7-9* and *7-10* show two forms of suitable RF detector probe.

The RF detector in *Figure 7-9* is *passive*; i.e., it has no amplification. It can be used around transmitters and other RF power sources. The input from the sensor if applied to C1, a small-value capacitor, and then is rectified by the 1N60 diode. The 1N60 is an old germanium type diode, and is used in preference to silicon diodes because it has a lower junction potential (so is more sensitive). The junction potential of Ge devices is 0.2 to 0.3 volts, and for silicon is 0.6 to 0.7 volts. The pulsating DC from the rectifier is filtered by capacitor C2. Resistor R1 forms a load for the diode and is not optional.

An amplifier version of the RF detector circuit is shown in *Figure 7-10*. In this version the same RF detector circuit is used, but it is preceded by a 15 to 20 dB gain amplifier. In this particular circuit the amplifier is a Mini-Circuits MAR-1 device. It produces gain from near-DC to about 1000 MHz. Other devices in the same series will work to 2000 MHz, and in the related ERA-x series up to 8000 MHz. Clearly, any of these devices is well suited to the needs of most readers. The cost of the MAR-1 device is very low. (In the USA it is about $3 in unit quantities.)

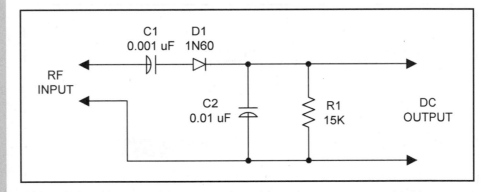

Figure 7-9.
RF detector converts
RF signal to DC.

Conclusion

Sleuthing out RF emitters that are causing problems can be a daunting task. But that task is made a lot easier by having some means for unambiguously locating the source of the offending RF signals. The sensors discussed in this chapter fit that bill. Once the source is found, a suitable strategy for eliminating the problem can be found.

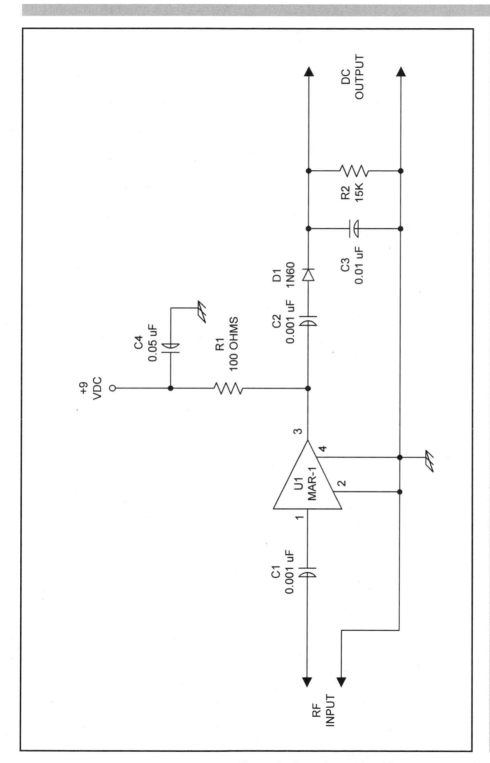

Figure 7-10.
Amplified RF
detector.

Chapter 8
MONITORING THE ULTRAVIOLET RADIATION FROM THE SUN

Chapter 8
Monitoring the Ultraviolet Radiation from the Sun

Forrest M. Mims III
Sun Photometer Atmospheric Network (SPAN)
(An Invited Chapter)

The regular measurement of the ultraviolet from the Sun that reaches the surface of the earth is an area in which the amateur can make significant contributions. This chapter describes the design, construction and operation of an instrument that can be used to monitor trends in the amplitude of solar ultraviolet radiation and the attenuating effects of clouds and air pollution. This instrument can also be used to measure the ultraviolet attenuation of sunscreening ointments, sunglasses and windows.

This chapter is a revised and updated version of an article originally published in The Amateur Scientist Department of *Scientific American* ("How to Monitor Ultraviolet Radiation from the Sun," Scientific American, 263, 2, 106-109, August 1990).

Ultraviolet Radiation

The Sun is a prodigious generator of ultraviolet radiation. While only a few percent of the radiation from the Sun is classified as ultraviolet (UV), the intensity of UV at the top of the atmosphere is more than adequate to preclude the existence of life as we know it. *Figure 8-1* shows the typical intensity of UV for three different latitudes at local solar noon when the Sun is at its highest point in the sky. Notice that the UV spectrum is divided into three sections: UV-A (315/320 to 400 nm), UV-B (280 to 315/320 nm) and UV-C (< 280 nm). UV-C radiation is extremely energetic and, therefore, quite dangerous to living cells. The surface of the planet is shielded from the damaging effects of UV-C by a thick but vacuous blanket of pale blue, toxic gas known as *ozone*. Excessive exposure to UV-B is also dangerous to life. Fortunately, the ozone layer blocks most of the UV-B. Enough leaks through to help sanitize the environment by killing bacteria and viruses and to produce vitamin D (good) and cancer (very bad) in human skin tissue. UV-B that causes a reddening of human skin is known as *erythemal radiation*.

The state of the ozone measurement art is such that a person equipped with a personal computer, a modem and Internet access can establish an indirect data link with the Total Ozone Mapping Spectrometer (TOMS) aboard NASA's current ozone satellite, at this writing the *EarthProbe* TOMS. The total ozone above any point on earth for the preceding day can be determined simply by going to the TOMS web site (http://toms.gsfc.nasa.gov) and entering the appropriate coordinates. For example, for my site in Texas I enter yesterday's date and the coordinates 29.6 (the latitude) and -97.9 (the longitude). Note that a longitude west of Greenwich is preceded by a negative sign. The site allows you to retrieve ozone values going back to November 1978.

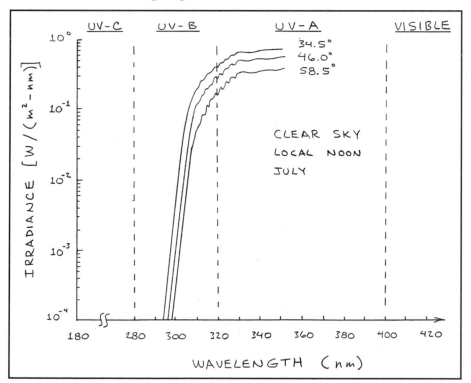

Figure 8-1.
Solar irradiance for three latitudes. (computed by John Frederick.)

When this chapter was first published as a column in *Scientific American* in August 1990, the only network of ultraviolet monitoring stations in the United States consisted of fewer that two dozen Robertson-Berger (R-B) meters. These devices have a broad spectral response designed to match the erythemal action spectrum, the band of ultraviolet wavelengths that causes reddening and eventual burning of human skin. The old network has been largely supplanted by a much bigger one operated by the Department of Agriculture. The Environmental Protection Agency operates a smaller network of more sophisticated UV meters. Several additional instruments are operated by various laboratories and at my site in Central Texas. Many countries have at least one UV monitoring site. The UV networks in Canada, Australia and New Zealand are among the world's best.

Both NASA and NOAA have developed algorithms that estimate the ultraviolet at the Earth's surface based on the amount of ozone and aerosols. NOAA UV forecast estimates of UV for 58 major U.S. cities are published at (http://nic.fb4.noaa.gov/products/stratosphere/uv_index/) on the Internet. NASA plans to publish much higher resolution UV estimates on its TOMS web site.

The old R-B meter network began operation in 1974. Its most important finding was that between 1974 and 1985 the average flux measured by eight meters in the network fell some 0.7 percent per year. Since stratospheric ozone has fallen by approximately 0.3 percent per year since 1978, an increase in ultraviolet would have been expected. The reason for the decrease is unclear, and some of the difference could be a result of calibration problems with the instruments. Other possible contributing factors are scattering and absorption of UV by meteorological phenomena and air pollution.

The R-B meters used for the 1974-85 study are all located in urban areas. An earlier comparison of R-B meters located in urban and nearby rural areas showed that the rural meters received from 5 to 7 percent more UV than the urban meters. Did urban air pollution skew the results of the 1974-85 study?

Here is where serious amateur UV-B observers can make a valuable contribution. Consider, for example, urban-rural pairs of schools, universities or individual observers equipped with identical UV-B meters. Over the course of a year, a comparison of simultaneous, daily UV-B observations would provide important information about the effect of urban air pollution on UV-B.

Global, Diffuse and Direct UV

Before describing the construction of a UV-B radiometer, it is important to define what is to be measured. As shown in *Figure 8-2*, the Sun and sky are distinctly separate sources of UV-B. Air molecules scatter UV-B radiation so efficiently that when the Sun is low in the sky most of the UV-B that arrives at a detector has been scattered from the sky. This gives rise to several important terms. *Global* or *total radiation* is the sum of the radiation direct from the Sun and scattered by the sky. *Diffuse radiation* is that component scattered by the sky, and *direct radiation* is that from the Sun alone.

The magnitude of global radiation is of high interest in studies of the deleterious effects of ultraviolet on both living systems and materials such as paints and plastics designed for outdoor applications. Global measurements are also important in studies of the effects of clouds on UV-B. The UV-B instruments used in various networks measure global UV-B.

Direct radiation measurements provide valuable information about the presence and effect of absorbing and scattering agents in the atmosphere. Because of the unpredictable nature of clouds and the possible presence of barriers such as buildings and trees, direct Sun measurements are preferred when the effect of air pollution on the relative magnitude of UV-B at two or more locations is to be compared.

So should the amateur observer monitor global or direct UV-B? Back in 1990 I discussed this question with Dr. John E. Frederick of the Department of Geophysical Sciences at the University of Chicago. Frederick has devised a computer model that predicts the levels of UV-B at Earth's

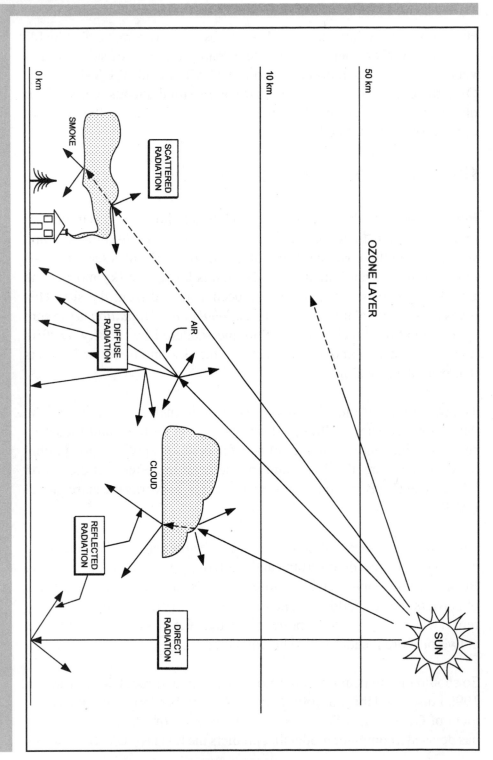

Figure 8-2.
Principal components
of Earth's ultraviolet
radiation budget.

surface for a range of conditions, and *Figure 8-1* is a product of this model. Frederick suggested that the amateur first concentrate on measuring direct UV-B. This will permit the comparison of measurements without being affected by the many variables that can impair global measurements. Therefore, the instrument described here is designed to detect direct UV-B. While I have since built several instruments that measure full-sky UV-B, I still monitor direct-Sun UV-B to watch various effects caused by aerosols and ozone.

UV-B Radiometer Basics

Since 1988, every day that the Sun shines I measure its radiation at several pairs of closely spaced wavelengths that permit the calculation of total water vapor, oxygen and ozone in a vertical column through the earth's atmosphere above my country office and shop in Central Texas. The most important of these measurements is at 308 (nm) in the UV-B. This wavelength is near the center of the portion of the UV-B spectrum that causes sunburn and DNA damage.

For these measurements I use a variety of homemade UV-B radiometers. The essential elements of these and other electronic UV-B monitoring instrument are a means for selecting the wavelength to be detected, a UV sensing detector, a low-noise direct-current amplifier and a digital voltmeter, or either an analog chart recorder or a computerized data acquisition system. Also required is some form of input optics. In the case of a direct detector, a simple collimator tube will suffice.

A monochrometer provides a convenient but expensive way to measure UV-B across a wide range of discrete wavelengths at a bandpass of 1 nm or less. An optical interference filter provides a much cheaper, simpler and more compact method for selecting a reasonably narrow band of UV-B wavelengths. Another important advantage of the filter approach is that it permits considerably more radiation to reach the detector. Chief among its disadvantages is that UV-B interference filters ordinarily have a bandpass of from 5 to 10 nm, considerably more than that of a monochrometer. Another drawback is that filter specifications can change with temperature and over time. Water vapor can permanently alter the bandpass of a filter and lead to the destruction of its laminated layers of thin films. Still another disadvantage of interference filters is that they transmit small but detectable levels of radiation outside their specified passbands. This disadvantage is often the most serious since it can cause significant errors.

A UV-B Radiometer

Figure 8-3A is the circuit diagram for a UV-B radiometer, with additional detail on the sensor in *Figure 8-3B*. I have used this circuit or variations of it in many different UV instruments since 1988. In operation, sunlight entering a collimator tube is filtered by a UV-B filter. The UV-B stimulates a photocurrent in photodiode PD1. This current is amplified and converted to a voltage by operational amplifier IC1. The amplified voltage is displayed on an Acculex DP-650 or similar digital panel meter.

A UV-enhanced silicon photodiode can be used for the photodiode, but it will also be very sensitive to unwanted red light that leaks through most UV filters. Gallium phosphide (GaP), silicon carbide (SiC) and gallium nitride (GaN) detectors that do not respond to red light will work much better. Several GaP diodes are made by Hamamatsu Corporation. The G1961 is housed in a miniature TO-18 package and has an effective surface area of 1.0 square millimeter. The G1962 is housed in a larger TO-5 package and has a surface area of 5.2 square millimeters. Other UV detectors can be found by searching Internet web sites.

Hamamatsu Corp.
P.O. Box 6910,
Bridgewater, NJ
08807

The most expensive component of the radiometer is the optical filter. The best filters I have used are made by Barr Associates. I use several different Barr UV-B filters with a 5-nm bandpass to monitor both solar UV-B and column ozone. Barr makes filters only on a custom basis; therefore, unless you are connected with an institution that can afford to place a custom order, you will need to go elsewhere. Relatively inexpensive UV filters are made by MicroCoatings and Twardy Technology Inc. You can find manufacturers of better quality filters on the web or in

Barr Associates P.O.
Box 557, Westford,
MA, 01886

MicroCoatings
One Lyberty Way,
Westford, MA 01886

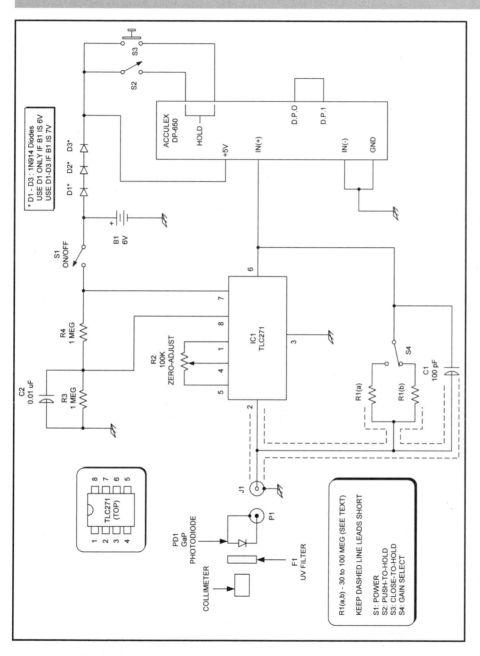

Figure 8-3A.
Circuit diagram for
one of the two
channels of the
TOPS ozonometer.

various trade magazines. If you want to measure the UV-B that causes sunburn, your best choice is a center wavelength of about 308 nm and a passband of 10 nm.

Twardy Technology
Inc.; P.O. Box 2221,
Darien, CT 06820

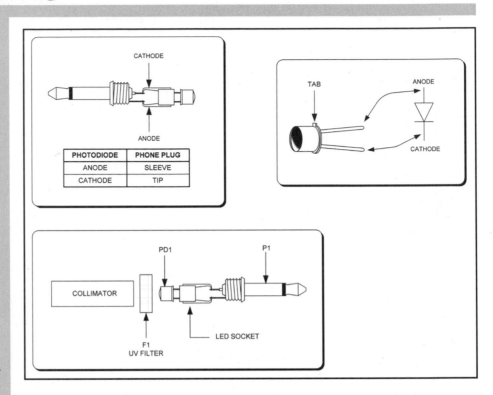

Figure 8-3B.
Additional detail on
the sensor assembly.

Use caution when buying UV-B filters. The best filters are generally the most expensive. But even expensive filters can have problems. I have worked with a few hundred UV-B filters made by several companies. Unfortunately, many filters transmit excessive out-of-band radiation, especially in the red and UV-A. Therefore, it is important to specify that you wish minimal out-of-band transmission. Also, be sure to select a filter that has some form of sealing around the edges to keep out water vapor. All UV-B filters will eventually degrade. Those without a seal will degrade much more rapidly.

Amplification of the radiometer is determined by feedback resistor R1, two values of which are provided in *Figure 8-3*. Switching between the two resistors allows two different amplification levels. You can omit R1(b) and S4 if you need only one level of amplification. It isn't possible to specify an exact value for R1 because changes in the sun's angle and different kinds of photodiodes and UV-B filters will provide very different photocurrents. Twenty to 30 megohms provides satisfactory resistance for R1(a) for a full year of measurements at my latitude (29° 35' N).

When using very high values of feedback resistance, considerable care must be exercised to preclude false signals caused by leakage currents between the detector's inverting input pin and ground. For example, a thin film of dust, moisture or oil may provide a path for an error generating current.

Ys are available from some electronics and surplus stores. Eltec Instruments, Inc. makes ultra-miniature, multiple-megohm resistors. Other makers of high-resistance resistors include the Robert G. Allen Company, RCD Components, Inc. and Victoreen, Inc.

Eltec Instruments, Inc.; P.O. Box 9610, Daytona Beach, FL 32020

Robert G. Allen Co. 7267 Coldwater Canyon, N. Hollywood, CA 91605

RCD Components, Inc.; 520 E. Industrial Park Dr., Manchester, NH 03103

Victoreen, Inc. 6000 Cochran Rd., Cleveland, OH 44139-3395

Texas Instruments P.O. Box 225012, Dallas, TX 75265

You can assemble your own high-resistance resistor by connecting in series the necessary number of 10-megohm resistors. Three 10-megohm resistors provide a 30-megohm resistor. These resistors are readily available from Radio Shack and most electronics parts stores.

The operational amplifier in *Figure 8-3* (TLC271CP) can be purchased from Texas Instruments sales offices and major electronics mail-order distributors. If you are unable to find a source locally, contact Texas Instruments and request a list of distributors and a data sheet for the TLC271CP. Many other op-amps can also be used if they have a very small input bias current. Be sure to check the manufacturer's data sheet for the op-amp you intend to use, since it may require a slightly different method for connecting R2.

The TLC271CP and some other op-amps with a very low input offset bias current can be damaged by static electricity. Therefore, avoid touching the pins of the op-amp while installing it. The best way to avoid this problem is to solder an 8-pin integrated circuit socket into the circuit. Install the op-amp after the circuit is assembled but before applying power.

The radiometer can be installed in a pocket-sized plastic enclosure. If you have had no prior experience assembling electronic circuits you should consider using a larger enclosure. In any event, the installation method is not critical. I installed the detector on one side of the perforated board that was supplied with the enclosure. If there is a chance light may leak through the enclosure, it might be necessary to carefully coat the back of the detector with black enamel. Do this if the back side of the detector shows any sensitivity to light. One path for light is through the insulated opening for one or both of the photodiode's pins.

Interconnections can be made with wirewrapping wire. It is important that the connections between the input of the amplifier, the photodiode and R1 be kept short and direct. Use a 9-volt battery clip to provide electrical contact with the batteries.

If you wish to save the expense of a built-in digital panel meter, include a pair of output leads equipped with pin jacks to receive the probes from a miniature digital multimeter.

The most important consideration of the radiometer is the installation of the detector and filter in a lighttight housing. *Figure 8-4* shows one method I have used that permits the detector-filter assembly to be quickly detached from the radiometer's enclosure.

The detector-filter assembly in *Figure 8-4* will hold a 12.5 mm (0.5 inch) filter and a small photodiode. It is made from a 3/8 inch brass compression fitting or union coupling. This coupling and the required O-rings are available from hardware and plumbing stores. A 1/8-inch phone plug is inserted in one of the union's caps and secured in place by means of a rubber O-ring. I soldered a light-emitting diode (LED) socket to the phone plug's terminals. You can, however, solder the detector directly to the terminals. The cathode lead should be soldered to the terminal that is common to the phone plug's tip. Depending on the detector's surface area, a conical endcap will give an FOV of around 10 degrees. Therefore you should install a collimator to reduce the FOV to at least 4 degrees or less. Brass tubing can be soldered or cemented to opening in the endcap. Be sure to coat the inside of the tube with flat black paint.

The filter, protected by a pair of O-rings, is installed in the second endcap. A conical cap works best but may be hard to find. If the filter and O-rings don't leave sufficient space for the endcap's threads to engage those of the union, replace one of the O-rings with a paper spacer. Screw the endcap down so that it stays in place without applying significant pressure on the filter. If necessary, use a drop of cement to keep the endcap from slipping loose.

It is essential that the filter remain clean during the installation procedure. Before installing the filter, be sure its surface is absolutely clean since UV-B is absorbed by dust, oil and other surface contaminants. Dust should be blown away with an air duster. Fingerprints and the like should be cleaned by gently swabbing the surface of the filter with ethyl alcohol. Never soak the filter in any liquid!

Between measurements I sometimes store my filter-detector assemblies along with a package or two of silica gel desiccant in an airtight plastic refrigerator container. This helps protect the filters from possible deterioration caused by long-term exposure to water vapor. Ask a pharmacist or salesperson at an electronics or camera store to save silica gel packets for your radiometer. I also use bulk silica gel intended for drying flowers, available from craft stores. Use silica gel with color indicators that are blue when dry and pink when moist. You can dry silica gel by placing it in an oven at 250 degrees F for a few hours.

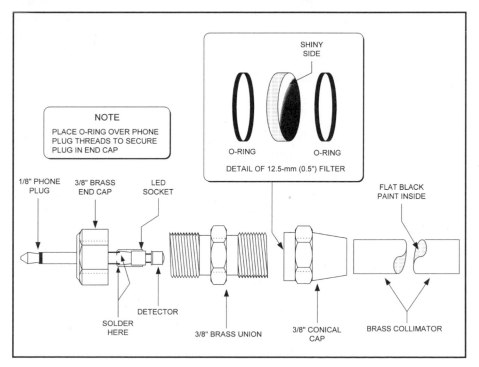

Figure 8-4. Installation of the detector and filter in a brass union coupling.

Using the Radiometer

The assembled radiometer is simple to use. First, carefully check the wiring for possible errors. After the voltmeter is connected (if you didn't include a built-in panel meter) and the power switch is toggled on, block the opening of the collimator tube and adjust R2 until the output voltage falls to zero. Be sure to repeat this procedure prior to each measurement session. Then point the collimator tube toward the Sun and align the tube until its shadow disappears. The detector will now have an unrestricted view of the Sun. Record the voltage, and make another measurement. You will soon find that even on a clear day the signal level fluctuates. The variations are considerably more pronounced near noon and when clouds, smoke or dust are present.

If the meter shows a negative signal, immediately switch off the power and recheck the wiring. If the meter shows a constantly changing signal even when the collimator tube is blocked, the amplifier may be detecting the electromagnetic radiation from a nearby power line. Try moving away from power lines or buildings. If this problem persists, it may be necessary to install the circuit in a metal enclosure. Connect the ground side of the op-amp input (pin 3) to the enclosure. Alternatively, you might be able to line the inside of a plastic enclosure with a shield of foil-backed tape. Use a small screw, nut and solder lug to connect the circuit's ground to the foil.

Your readings may include an error factor contributed by the detector's response to the red light that leaks through the UV-B filter. The simplest way to eliminate this error is to follow each reading with a second reading, during which you place a UV blocking filter over the entrance of the collimator tube. A UV filter intended for a camera works well, as does a WG-345 clear glass filter. If the meter indicates a signal of 0 when the radiometer is pointed directly at the Sun through a UV blocking filter, the second reading is unnecessary. Subtracting the second reading (B) from the first (A) will give you a voltage which will be linear with respect to measurements you make at other times.

By now you may be wondering how the radiometer is calibrated. Unfortunately, calibration requires use of an expensive standard lamp or a comparison with a professional UV-B Sun photometer (not a full-sky instrument). Some universities and government labs have the means to calibrate the radiometer, and a polite request might yield assistance in this regard. If you are unable to calibrate your detector, you can still use the instrument to track changes in UV-B. This is possible since the response of the instrument is linear with respect to UV.

If you have a calibrated detector, you can compute the absolute spectral irradiance at your filter's wavelength in terms of watts per square meter. Each surface of the UV blocking filter reflects around 4 percent of the oncoming radiation for an overall transmission of 92 percent. Therefore, if you use the blocking filter for a second reading, you will need to divide the B reading by 0.92. The active area of one photodiode I have used is 9.9 square millimeters; therefore, the detector signal must be multiplied by 101,010 to find the signal per square meter. The equation that results is $(101{,}010 \times (((A-(B/.92))/R1)/Dr))/F$, where Dr is the detector's calibrated responsivity and F is the filter's bandpass. My detector has a Dr of .04 amperes per watt and a filter bandpass of 10.4 nm at the half amplitude points. A typical pair of clear-day readings at noon during August is 1.501 volts (A) and .1155 (B) volt. Inserting these values into the formula gives $(101{,}010 \times (((1.501-(.1155/.92))/30{,}000{,}000)/.04))/$ 10.8 or 0.0107 watts/square meter-nm. The diffuse contribution from radiation scattered by molecules in the atmosphere (Rayleigh scattering) would have added at least 30 percent to this value at my latitude.

A programmable calculator greatly simplifies the application of the formula. Here, for example, is how the formula appears when converted into a program for a Hewlett-Packard programmable scientific calculator: '101010∗(A-B/.92)/30000000/.04/10.8'.

If you prefer to use a computer, a simple BASIC program can be written to solve the formula. An even better approach is to program a computer spreadsheet to both solve the formula and save and tabulate your measurements along with the date, time and other parameters. I have used all three methods, and the last is by far the best.

Significant knowledge about direct solar UV-B can be obtained with a program of regular measurements using either a calibrated or uncalibrated detector. Ordinarily I make measurements at local apparent noon in order to measure the day's peak signal. Standard time is converted to apparent time by adding 4 minutes for each degree of longitude west of the standard meridian. In the United States, the standard meridians for Eastern, Central, Mountain and Pacific time are, respectfully, 75, 90, 105 and 120 degrees. The equation of time causes solar noon to arrive as much as some 16 minutes early or 14 minutes late. For a precise knowledge of solar noon, consult any standard reference on astronomy or Sundial construction. Information on this topic can also be found on the web.

Some Results

The peak UV-B reading will occur at or near solar noon. *Figure 8-5* shows the direct UV-B for two entire days, one near the summer solstice and one near the fall equinox. The signal fluctuates constantly as the UV-B radiation is attenuated and scattered by the atmosphere and its constituent gases and aerosols.

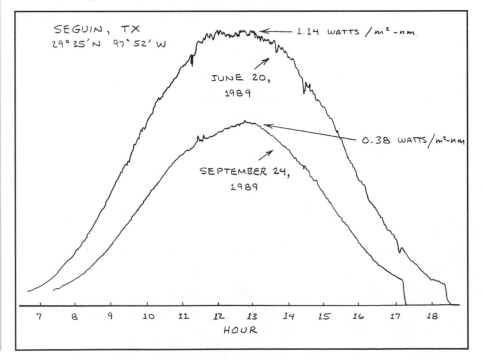

Figure 8-5.
Direct UV-B flux
(300 nm) near
Summer Solstice
and Autumnal
Equinox.

My daily program of measurements has revealed that direct solar UV-B irradiance is significantly attenuated by fog, haze, clouds and high altitude aircraft contrails which have evolved into cirrus clouds. During the Yellowstone fires in the summer of 1988, I observed major decreases in UV-B on two occasions when smoke drifted over Texas (*Figure 8-6*).

I have also learned that the passage of a cold front is more often that not followed by a reduction in UV-B, even when the postfrontal passage atmosphere is exceptionally clear and dry. Thanks to measurements made by NASA's TOMS, I was able to confirm that this phenomenon is a direct result of an increase in ozone. I have since made many such observations of UV-B decreases following cold fronts and have confirmed simultaneous ozone increases by accessing NASA's TOMS database and by using my own ozone instruments.

If cold fronts are followed by increases in both barometric pressure and ozone, low pressure, at least where I reside in central Texas, is typically

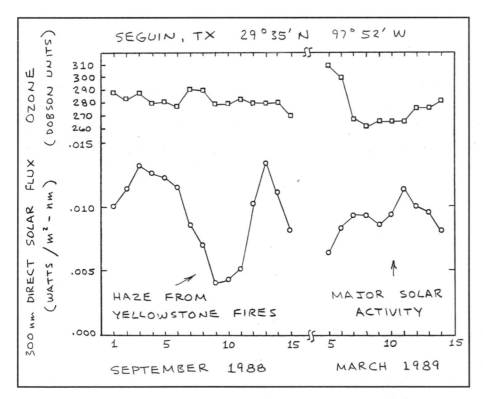

Figure 8-6. Observations of 300 nm solar flux during the Yellowstone fires and an episode of major solar activity.

accompanied by a decrease in ozone. A particularly dramatic example of this occurred when Hurricane Gilbert passed near my home during September 16-18, 1988. My radiometer revealed a significant increase in UV-B during this time. TOMS data revealed a simultaneous and pronounced decrease in ozone.

Sometimes I mount a UV-B radiometer on a clock-driven telescope mount to track the Sun for days at a time. The instrument is connected to a data logger or chart recorder. This method provides very interesting data about the effects of Sun angle, seasons and clouds on direct UV-B. However, prolonged exposure to sunlight will eventually damage UV-B filters.

A Commercial Global UV-B Radiometer

Dozens of people assembled the radiometer described here after its original publication in *Scientific American.* While constructing your own UV-B radiometer is both challenging and fulfilling, you might wish to consider buying one of the new commercial UV-B radiometers. These instruments sell for about the same price as the cost of individual components for a do-it-yourself version.

Solartech Inc.
37512 Jefferson,
Suite 103, Harrison
TWP, MI 48045;
800-798-3311

I have extensively tested economical UV-B radiometers made by three companies. By far the best of these instruments is made by Solartech Inc. Solartech's Model 5.0 Solarmeter measures global (full-sky) total UV. It is equipped with a built-in digital panel meter. I have tested two of these instruments for two years against one another and a variety of professional instruments and my own radiometers here in Texas and at the Mauna Loa Observatory in Hawaii. The Solartech radiometers are very stable and, as of this writing, there has been no significant drift in their calibrations. Solartech also makes the Model 6.0 (UV-B) and the Model 6.5 (UV-B index).

The Solartech radiometers are designed for full-sky measurements. They can also measure direct-Sun UV if you place a temporary collimator tube over the detector.

[Note: Forrest Mims III is a well-known and highly respected technical writer and amateur scientist. He is a winner of the prestigious Rolex Prize given to outstanding amateur scientists. His work appears not only in electronics publications, but in scientific journals around the world.]

Chapter 9
SMALL RF COMPONENTS USED IN RADIOSCIENCE OBSERVING SYSTEMS

Chapter 9
Small RF Components Used in RadioScience Observing Systems

When you assemble a radio astronomy system, or other RadioScience Observing receiving system, you might want to consider some of the devices described in this chapter as part of the system. They make some jobs easier, and some jobs possible.

Resistive Attenuators

An *attenuator* is a device that reduces the amplitude of an applied signal. *Figure 9-1A* shows an attenuator in block form. A real attenuator may or may not have a ground connection depending on whether it is *balanced* (no ground) or *single-ended* (grounded). Examples of both balanced and unbalanced types are shown below. Most commercially available attenuators are shielded, and the shield will be grounded even if the circuit inside is balanced.

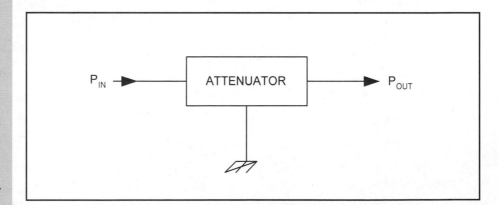

Figure 9-1A.
Symbol for attenuator.

Attenuation Definition. In an attenuator $P_{OUT} < P_{IN}$ by definition. The amount of attenuation can be expressed in either linear terms or in decibel notation, with the latter being more common. The attenuation factor in dB is:

$$Attenuation\,(dB) \;\; = \;\; 10\,Log\left[\frac{P_{OUT}}{P_{IN}}\right] \qquad\qquad (9\text{-}1)$$

Where:

Attenuation (dB) is the reduction of input signal in decibels

P_{OUT} is the output signal power level

P_{IN} is the input signal power level

Log refers to the base-10 logarithms

Note: P_{IN} and P_{OUT} are expressed in the same units (watts, milliwatts, microwatts).

Equation 9-1 gives the attenuation in terms of signal power level. To find the attenuation expressed in terms of input and output voltages or currents, replace the factor "10" with "20" in the equation. If you have the attenuation figure in dB, then the assumption is that it is relative to power levels. To find the voltage attenuation knowing the power attenuation, simply multiply power dB by 2; thus, a -3 dB attenuator will also have a -6 dB attenuation of voltage.

Impedances. The input and output impedances of attenuators of the attenuator are critical. Although circuits exist in which $Z_{IN} \neq Z_{OUT}$, the normal case is that $Z_{IN} = Z_{OUT}$. In most RF systems $Z_{IN} = Z_{OUT} = 50\Omega$, but in television and video work the usual rule is $Z_{IN} = Z_{OUT} = 75\Omega$ for unbalanced applications and 300Ω for balanced cases. When using an attenuator to perform tests and measurements it is necessary to use a unit of the correct input/output impedance. Otherwise, reflections will occur because of the impedance mismatch, and the marked attenuation factor is incorrect. In the case of an impedance mismatch there will be a loss based on the VSWR, which in turn is the ratio of the impedances. For example, if you connect a 50Ω attenuator into a 75Ω circuit, the VSWR will be 75Ω/50Ω = 1.5:1.

Attenuator Examples. *Figures 9-1B* through *9-1D* are examples of un-balanced fixed attenuators. The circuit in *Figure 9-1B* is a pi-network attenuator (named after the similarity to the Greek letter π). This circuit is probably the most common. *Figure 9-1C* is a Tee-network attenuator, while *Figure 9-1D* is a somewhat more complex attenuator circuit.

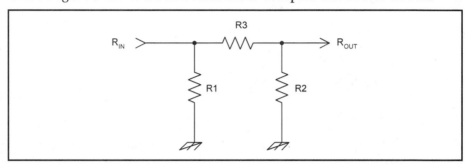

Figure 9-1B.
Pi-attenuator.

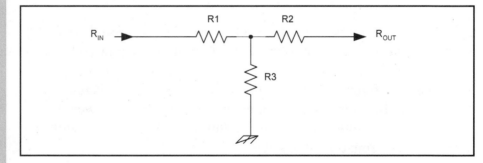

Figure 9-1C.
Tee-attenuator.

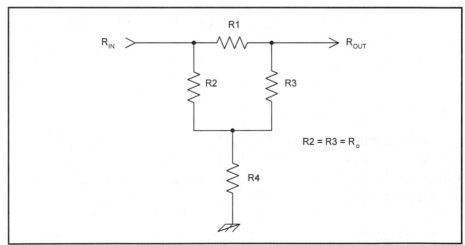

Figure 9-1D.
Attenuator network.

Table 9-1 shows the common values for the resistors in the Pi-attenuator of *Figure 9-1B*. In some cases, you might wish to build an attenuator, but normally that is not a wise use of time, especially given the fact that the resistor values are so specific. Commercial attenuators are available at very low cost from companies such as *Mini-Circuit Laboratories*.

Attenuation (dB)	R1 (Ohms)	R2 (Ohms)	R3 (Ohms)
1	870	870	5.8
2	436	436	11.6
3	292	292	17.6
6	150.5	150.5	37.3
10	96.2	96.2	71.2
12	83.5	83.5	93.2
20	61	61	247.5
30	53.2	53.2	789.7
40	51	51	2500
50	50.3	50.3	7905.6
60	50.1	50.1	25,000.00

Table 9-1.
Resistor values for various pi-attenuator networks.

Figure 9-2 shows the equivalent fixed attenuator circuits for balanced use. The version in *Figure 9-2A* is the balanced pi-attenuator, while that in *Figure 9-2B* is an H-pad attenuator. These circuits are used with balanced applications such as some television receivers. In the case of TV receivers, the balanced antenna scheme is normally 300Ω rather than 75Ω.

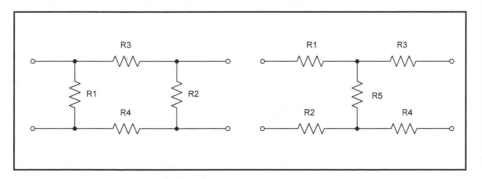

Figure 9-2.
A) Balanced pi-attenuator;
B) H-pad attenuator.

Switchable Attenuators. It is sometimes necessary to be able to switch attenuators in and out of a circuit. *Figure 9-3* shows a typical circuit in which a DPDT toggle switch is used to select either the attenuator or the pass-through path. Note that there is a shield around the circuit. It is very important to include the shield in order to prevent leakage around the attenuator from creating levels that are not accounted for by the measuring process. Note that not all switches are well shielded, so be careful when selecting a switch type for S1.

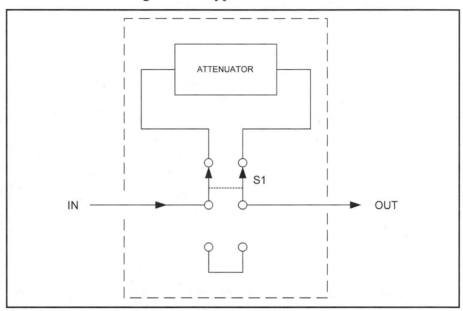

Figure 9-3.
Switchable
attenuator.

Laboratory Attenuators. It is unlikely that building an attenuator is a worthwhile use of your time and resources. *Figure 9-4* shows several types of commercial attenuators used in the laboratory. A fixed attenuator is shown in *Figure 9-4A*. This device is sometimes called a *barrel attenuator* or *in-line attenuator*. It has a male BNC on one end and a female BNC on the other. (Other connector combinations are also available, but BNC is the most common.)

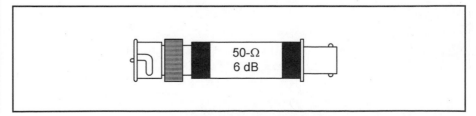

Figure 9-4A.
Fixed "barrel" or "in-line" attenuation.

A *step attenuator* is shown in *Figure 9-4B*. It consists of a series of circuits such as *Figure 9-3*, each in its own shielded compartment within the shielded enclosure for the entire device. In this version female BNC connectors are on either end, while the toggle switches are on the top. To select an amount of attenuation the required switches are turned on or off as needed. For example, to make a 3-dB attenuator turn on the 1-dB and 9-dB switches, and leave all others turned off.

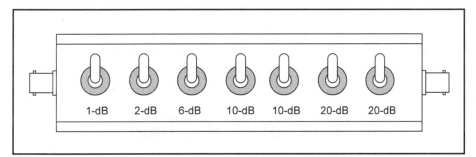

Figure 9-4B. Step-attenuator.

An electrically switchable step attenuator is shown in *Figure 9-4C*. This device has feedthrough or EMI filtering capacitors mounted on the shielded enclosure. Applying a DC voltage to the connector on the capacitor turns on the associated attenuator. In some cases, electronic switching is used inside the box, but in others electromechanical relays are used.

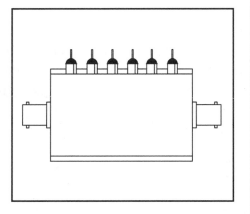

Figure 9-4C. Electrically switched attenuator.

There are other forms of attenuator used, but these are useful only at the VHF and below range. *Figure 9-5A* shows an inductive attenuator. It consists of two coils, one of which is station and the other rotates within the first. When the two coil axes are aligned, coupling is maximum so the signal output is highest. But when

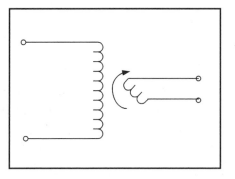

Figure 9-5A. Inductive attenuator.

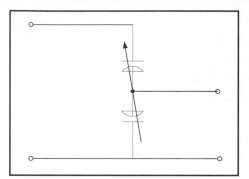

Figure 9-5B.
Capacitive
attenuator.

they are orthogonal the output is zero. The other approach is the capacitive voltage divider in *Figure 9-5B*. It uses a two-section differential capacitor as a voltage divider. A differential capacitor is one that has two identical sections mechanically linked 180 degrees out of phase with each other; in other words, as one increases in value the other decreases by the same amount. The total capacitance across the series pair is constant, but their ratio is not.

BALUN and Impedance Matching Transformers

There are at least two reasons why you might want to use a transformer in an RF test or measurement setup. One is to match impedances between two different devices. Another is to convert between balanced and unbalanced circuits. In the latter case, a BALUN (BALanced-UNbalanced) transformer is used.

Figure 9-6 shows several different transformer styles. The transformer shown in *Figure 9-6A* is a BALUN with a 1:1 impedance ratio. It is used to provide translation between balanced and unbalanced circuits, but without impedance transformation. The version in *Figure 9-6B* is also a BALUN transformer, but provides a 4:1 impedance transformation.

The transformers in *Figures 9-6C* and *9-6D* are both unbalanced on both ends (some call them "UN-UNs"). The transformer of *Figure 9-6C* provides a 9:1 impedance transformation, while *Figure 9-6D* provides 16:1 transformation. These circuits are predicated on using an equal number of turns on each winding. Other transformation ratios can be created using different turns ratios, or by tapping a single winding.

Figure 9-6E shows how these transformers are constructed. The wire is wound on a ferrite or powdered iron toroidal shaped core using either the *bifilar* (two windings, as shown in *Figure 9-6E*), *trifilar* (three windings) or *multifilar* (any number of windings) approach.

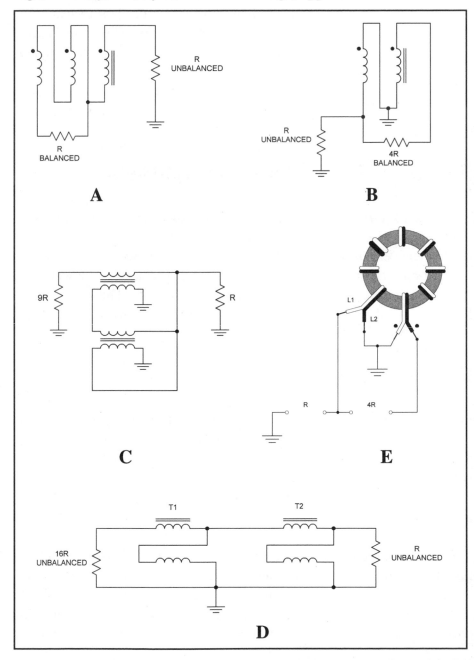

Figure 9-6.
Transformer circuits:
A) 1:1 BALUN;
B) 4:1 BALUN;
C) 9:1 transformer;
D) 16:1 transformer;
E) bifilar winding.

Coaxial Switches

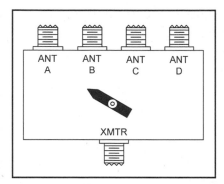

Figure 9-7.
Coaxial switch.

If it is necessary to switch between devices, receivers, antennas and/or loads, then it might be prudent to use a *Y*, such as the example shown in *Figure 9-7*. These devices are usually designed for amateur radio operators, so the labeling reflects that application. However, it is also useful for other RF switching purposes.

RF Power Combiners and Splitters

The principal difference between power combiners and power splitters is in the application. Otherwise, they are the same circuits. A *combiner* is used whenever it is necessary to linearly mix two or more signal sources into a common port. The combiner is not a mixer because it is linear, and thus does not produce additional frequency products. The *splitter* performs exactly the opposite function. It will direct RF power from a single source to two more loads.

Resistive Combiner/Splitter

Perhaps the simplest circuit is the resistive network in *Figure 9-8*. This circuit uses three resistors in a Y-network to provide three ports (it can also be extended to higher numbers of ports). The value of each resistor is $R = R_o/N$, where R_o is the system impedance and N is the number of ports. For example, if the system impedance is 50 ohms, then R = 16.67 ohms, and for 75-ohm systems R = 25 ohms.

The resistors used in this circuit must be *noninductive*. This limits selection to carbon composition or metal film resistors. If higher power than 9 watts is needed, then each arm of the Y-network must be made from multiple resistors in series or parallel. The values 16.67 ohms and 25 ohms are not standard values, except perhaps in certain lines of 1% or

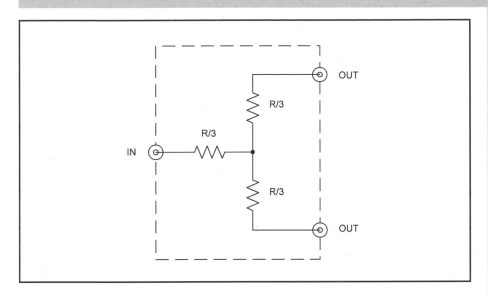

Figure 9-8.
Resistive combiner/
splitter.

less tolerance precision resistors. They can, however, be approximated using standard values. For example, only a small error is created when 15-ohm resistors are used in place of 16.67 ohms, and 27 ohms is used in place of 25 ohms. Because resistors come with variation in actual value, the amount of which is indicated by its tolerance rating (5%, 10%, 20%), we can often select from a collection of standard values to closely approximate the actual value needed.

It is also possible to approximate the values by using series or parallel combinations of standard-value resistors. For example, a pair of 51-ohm standard-value resistors in parallel will make a good match for 25 ohms. Similarly, three 51-ohm resistors in parallel will closely approximate 16.67 ohms.

The advantages of the resistive combiner/splitter is its broadband operation. The bandwidth can extend into the UHF region with discrete resistors, and into the gigahertz region if implemented with surface-mount resistors and appropriate printed-circuit technology. The upper frequency limit in either case is set by the stray inductance and capacitance.

The disadvantages of this form of combiner include a relatively high insertion loss, 6 dB, of which 3 dB is due to the resistors, the other 3 dB due to the fact that the input power is split two ways. Isolation between output ports is about 6 dB. If those can be overcome, or are not important in a given application, then this form of splitter/combiner is ideally suited.

Transformer Combiner/Splitter

Figure 9-9 shows a somewhat better form of combiner/splitter circuit. This circuit can be used from 500 kHz to over 1,000 MHz if the proper transformers and capacitor are provided. In this discussion let's concentrate on the high-frequency shortwave bands, as those are the easiest form of combiner/splitter for most readers to actually build.

The power splitting function is performed by coil L2. This coil is center tapped, with the input signal applied to the tap and the outputs taken from the ends. This transformer can be wound on either T-50-2 or T-50-6 toroidal cores for the HF bands, or a T-50-15 core for the AM BCB and medium wave bands. Use 18 turns of #26 AWG wire for the HF bands, and 22 turns for MW bands.

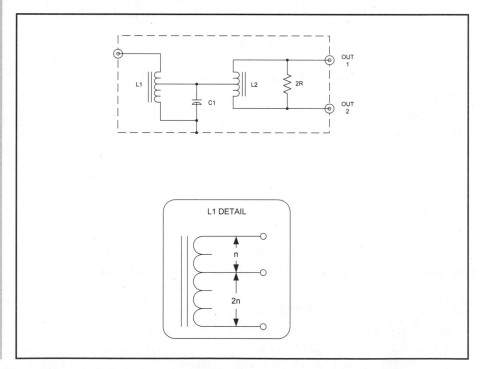

Figure 9-9.
Transformer
combiner/splitter.

The resistor across the ends of L2 should be twice the system impedance; that means 100 ohms for 50-ohm systems, and 150 ohms for 75-ohm systems (both are standard values).

Some impedance transformation is needed if the system impedance is to be maintained, so L1 must be provided. This transformer is tapped, but not at the center. The inset detail in *Figure 9-9* shows the relationship of the tap to the winding: it is located at the one-third point on the winding. If the bottom of the coil is grounded, then the tap is at the two-thirds point (2N turns), and the input is at the top (N + 2N turns). In other words, the tap is at two-thirds the overall length of the winding.

The capacitor usually has a value of 10 pF, although people with either a sweep generator, or a CW RF signal generator and a lot more patience than I've got, can optimize performance by replacing it with a 15-pF trimmer capacitor. Adjust the trimmer for flattest response across the entire band.

It is important to use toroid core inductors for the combiner. The most useful core types are listed above, although for other applications other cores could be used. *Figure 9-10* shows one way the cores can be wound. This is the linear winding approach; i.e., uses a single coil of wire. The turns are wound until the point where the tap occurs. At that point one of two approaches is taken: 1) End the first half of the winding and cut the wire. Adjacent to the tap start the second half of the winding. Scrape the insulation off the ends at the tap, and then twist the two ends together to form the tap; 2) Loop the wire (see detail inset to *Figure 9-10*), and then con-

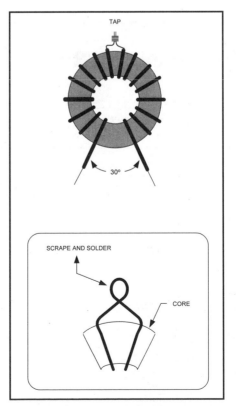

Figure 9-10.
Toroid winding.

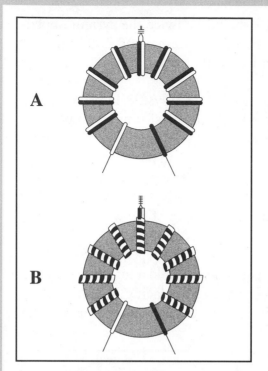

Figure 9-11.
Two different bifilar
winding styles.

tinue the winding. The loop then becomes the tap. Scrape the insulation off the wire, and solder it. Although the tap here is a center tap (which means L2), it also serves for L1 if you offset the tap a bit to the left or right.

An alternate method for L2 is shown in *Figure 9-11*. This is superior to the other form for L2, but it is a little more difficult. Either wind the two wires together side-by-side (*Figure 9-11A*), or twist them together before winding (*Figure 9-11B*). Make a loop at the center-tap, and scrape it for soldering.

Modified VSWR Bridge Combiner/Splitter

Figures 9-12 and *9-13* show 6 dB combiner/splitter circuits based on the popular bridge used to measure voltage standing-wave ratio (VSWR).

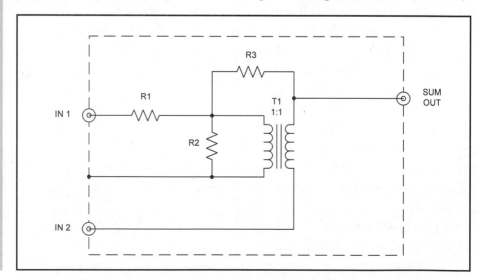

Figure 9-12.
VSWR bridge
combiner/splitter.

Each of these circuits use a bridge made of three resistors and one winding of a transformer. In both cases, the transformers have a 1:1 turns ratio. Also, in both cases $R1 = R2 = R3 = R_o$. In 50-ohm systems, therefore, the value of the resistors is 50 ohms, and in 75-ohm systems it is 75 ohms. In *Figure 9-12* the transformer is not tapped. It is a straight 1:1 turns ratio toroid transformer. The circuit in *Figure 9-13*, however, uses a center tap on the primary of T1. Note that there is a difference in the location of the summation output between the two circuits. These circuits have been popular for combining two signal generators; for example, a sweep generator and a marker generator.

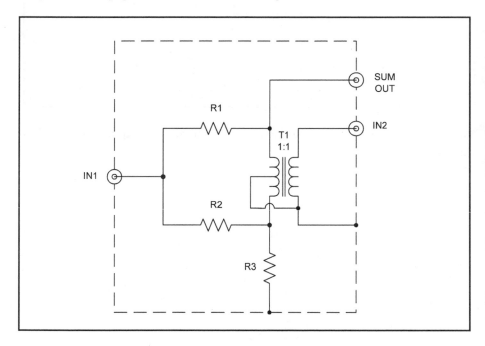

Figure 9-13.
Alternate VSWR
combiner/splitter.

A variation on the theme is shown in *Figure 9-14*. This circuit is sometimes also used as a directional coupler. RF power applied to the input port appears at OUT-2 with only an insertion loss attenuation. A sample of the input signal appears at OUT-1. Alternatively, if RF power is applied to OUT-2, it will appear at the input, but does not appear at the OUT-1 port due to cancellation.

For the case where the device has a -3.3 dB output at OUT-2 and a -10 dB output at OUT-1, and a 50-ohm system impedance (R_o), the

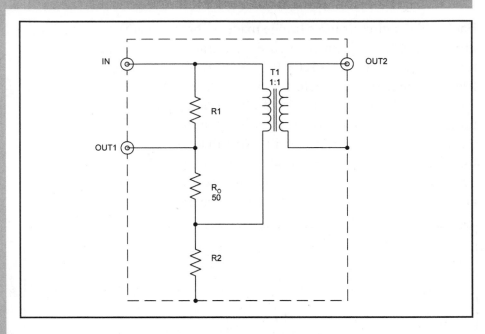

Figure 9-14.
Unequal output
combiner/splitter.

value of R1 = 108 ohms, and R2 = 23 ohms. The equations for this device are:

$$R_o \ = \ \sqrt{R1 \, R2} \tag{9-2}$$

$$C.F. \ = \ 20 \, Log\left(\frac{R_o}{R1 + R_o}\right) \tag{9-3}$$

$$L_i \ = \ -20 \, Log\left(\frac{R_o}{R2 + R_o}\right) \tag{9-4}$$

Where:

R$_o$ is the system impedance (e.g., 50Ω)

C.F. is the coupling factor

L$_i$ is the insertion loss from IN to OUT-2

R1 and R2 are the resistances of R1 and R2

Taking the negative of Equation 9-2 gives the insertion loss from IN to OUT-1.

90° Splitter/Combiner

Figure 9-15 shows a 3 dB splitter/combiner made of lumped L and C elements, and which produces a 0 degree output at OUT-1, and a 90 degree output at OUT-2. A closely coupled 1:1 transformer is used to supply two inductances, L1 and L2. This transformer is wound in the bifilar manner to ensure tight coupling. The values of inductance and capacitance, assuming that L1 = L2 = L, and C1 = C2 = C, are given by:

$$L = \frac{R_o}{2.828\,\pi\,f_{3dB}} \tag{9-5}$$

$$C = \frac{1}{2.828\,\pi\,f_{3dB}\,R_o} \tag{9-6}$$

Where:

L is the inductance of L1 and L2

C is the capacitance of C1 and C2

R_o is the system impedance (e.g., 50Ω)

f_{3dB} is the 3 dB coupling frequency

The bandwidth of this circuit is approximately 20 percent for 1 dB amplitude balance.

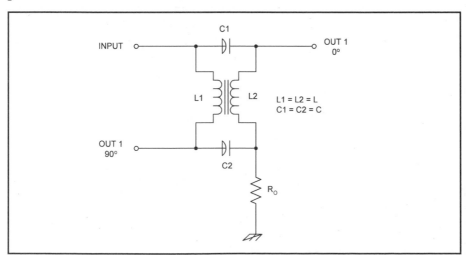

Figure 9-15.
Quadrature
combiner/splitter.

Transmission Line Splitter/Combiners

The Wilkinson power splitter/combiner is shown in *Figure 9-16*. This network can achieve 9-dB isolation between the two output ports over a bandwidth that is approximately ±20 percent of the design frequency. It consists of two transmission lines, TL1 and TL2, and a bridging resistor (R), which has a value of $R = 2R_o = (2)(50\Omega) = 100\Omega$.

Transmission lines TL1 and TL2 are each quarter wavelength, and have a characteristic impedance equal to 1.414 times the system impedance. If the system impedance is 50 ohms, then the value of the characteristic impedance needed for the transmission lines is $(1.414)(50\Omega) = 70.7\Omega$.

The Wilkinson network can be implemented using coaxial cable at VHF and below, although at higher frequencies printed-circuit transmission line segments are required. If coaxial cable is used, then the physical length of TL1 and TL2 are shortened by the *velocity factor* (VF) of the cable used for TL1 and TL2. The values of VF will be 0.66 for polyethylene dielectric coax, 0.80 for polyfoam dielectric, and 0.70 for *Teflon*™ dielectric cable. The physical length is:

$$Length \;=\; \frac{75\,VF}{F_{MHz}} \quad \text{meters} \tag{9-7}$$

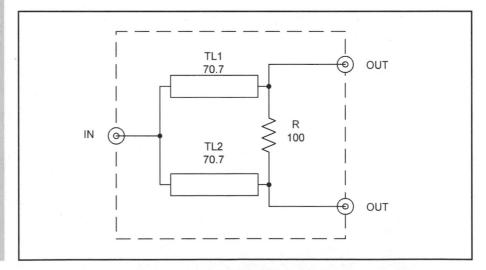

Figure 9-16.
Wilkinson
combiner/splitter.

An N-way version of the same idea is shown in *Figure 9-17*. In this network a transmission line, TL1 - TL(n), and resistor are used in each branch. The resistor values are the value of R_o. In the case shown, the values of resistors are 50 ohms because it is designed for standard 50-ohm systems. The characteristic impedance of the transmission lines used in the network is:

$$Z_o = R_o \sqrt{N} \qquad (9\text{-}8)$$

Where:

   Z_o is the characteristic impedance of the transmission lines

   R_o is the system standard impedance

   N is the number of branches

In the case of a 50-ohm system with three branches, the characteristic impedance of the lines is $(50\Omega)(\sqrt{3}) = (50\Omega)(1.73) = 86.5\Omega$.

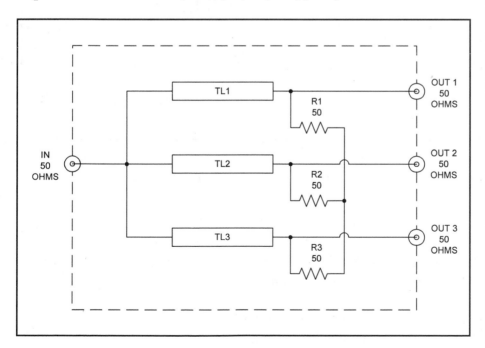

Figure 9-17.
N-way combiner/
splitter.

90° Transmission Line Splitter/Combiner

Figure 9-18 shows the network for producing 0°-90° outputs, with -3 dB loss, using transmission line elements. The terminating resistor at one node of the bridge is the system impedance, R_o (e.g., 50 ohms).

Each transmission line segment is quarter wavelength ($\lambda/4$), so have physical lengths calculated from Equation 9-7 above. The characteristic impedance of TL1 and TL2 is the system impedance, R_o, while the impedance pF TL3 and TL4 are $0.707R_o$. In the case of 50-ohm systems, the impedance of TL3 and TL4 is 35 ohms.

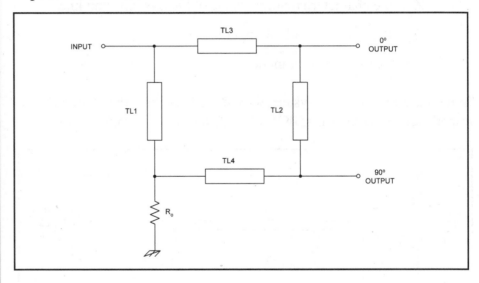

Figure 9-18. Transmission line quadrature combiner/splitter.

Hybrid Ring "Rat-Race" Network

The "Rat Race" network of *Figure 9-19* has a number of applications in communications. It consists of five transmission line segments, TL1 through TL5. At VUF, UHF and microwave frequencies this form is often implemented in printed-circuit board transmission lines.

Four of these transmission line segments (TL1-TL4) are quarter wavelength, while TL5 is half wavelength. The characteristic impedance of all lines is $1.414R_o$. Each quarter wavelength segment creates a 90-degree

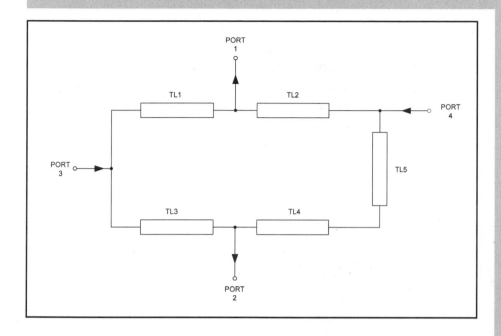

Figure 9-19.
Rat-race combiner/
splitter.

phase shift, while the half wavelength produces a 180-degree phase shift. When a quarter wavelength and half wavelength are combined, to form a three-quarter wavelength segment, the phase shift is 270 degrees.

It is necessary to terminate all ports of the Rat Race network in the system characteristic impedance, R_o, whether they are used or not. The bandwidth of this network is approximately 20 percent.

Different applications use different ports for input and output. *Table 9-2* shows some of the relationships found in this network.

Input	Use
Port 3	0° Splitter, -3 dB at Ports 1 and 2.
Port 4	180° Splitter, -3 dB, -90° at Port 1, -270° at Port 2.

Table 9-2.
Rat-race network
port uses.

A coaxial cable version is shown in *Figure 9-20*. This network is implemented using coaxial cable sections and Tee connectors. In this case, there are three quarter wavelength sections (90°) and one three-quarter wavelength (270°) section. Applications of this network include those where a high degree of isolation is required between ports.

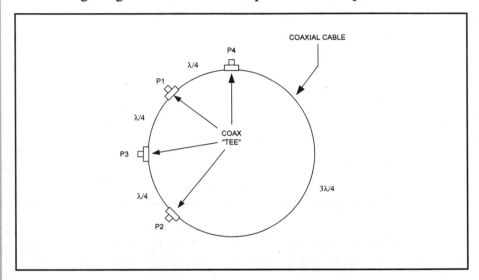

Figure 9-20.
Coaxial rat-race
combiner/splitter.

The RF Hybrid Coupler

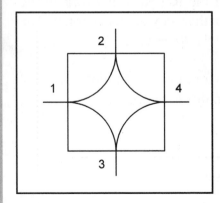

Figure 9-21.
Symbol for hybrid.

The hybrid coupler (*Figure 9-21*) is a device that will either split a signal source into two directions, or combine two signals sources into a common path. The circuit symbol shown in *Figure 9-21* is essentially a signal path schematic. Consider the situation where an RF signal is applied to Port-1. This signal is divided equally, flowing to both Port-2 and Port-3.

Because the power is divided equally the hybrid is called a 3 dB divider; i.e., the power level at each adjacent port is one-half (-3 dB) of the power applied to the input port.

If the ports are properly terminated in the system impedance, then all power is absorbed in the loads connected to the ports adjacent to the injection port. None travels to the opposite port. The termination of the opposite port is required, but it does not dissipate power because the power level is zero.

The one general rule to remember about hybrids is that *opposite ports cancel*. That is, power applied to one port in a properly terminated hybrid will not appear at the opposite port. In the case cited above, the power was applied to Port-1, so no power appeared at Port-4.

One of the incredibly useful features of the hybrid is that it accomplishes this task while allowing all devices connected to it to see the system impedance, R_o. For example, if the output impedance of the signal source connected to Port-1 is 50 ohms, the loads of Port-2 and Port-3 are 50 ohms, and the dummy load attached to Port-4 is 50 ohms, then all devices are either looking into, or driven by, the 50-ohm system impedance.

One source of reasonably priced hybrid devices is Mini-Circuits Laboratories. They have a large selection of 0-degree, 90-degree and 180-degree hybrid combiners and splitters.

Mini-Circuits Labs
13 Neptune Avenue,
Brooklyn, NY, 11235;
www.minicircuits.com

Applications of Hybrids

The hybrid can be used for a variety of applications where either combining or splitting signals is required.

Combining Signal Sources. In *Figure 9-22* there are two signal generators connected to opposite ports of a hybrid (Port-2 and Port-3). Power at Port-2 from Signal Generator No. 1 is therefore canceled at Port-3, and power from Signal Generator No. 2 (Port-3) is canceled at Port-2. Therefore, the signals from the two signal generators will not interfere with each other.

In both cases, the power splits two ways. For example, the power from Signal Generator No. 1 flows into Port-2 and splits two ways. Half of it (3 dB) flows the path from Port-2 to Port-1, while the other half flows from Port-2 to Port-4, similarly with the power from Signal Generator No. 2 applied to Port-3. It splits into two equal portions, with one flowing to Port-1 and the device under test, and half flowing to the dummy load.

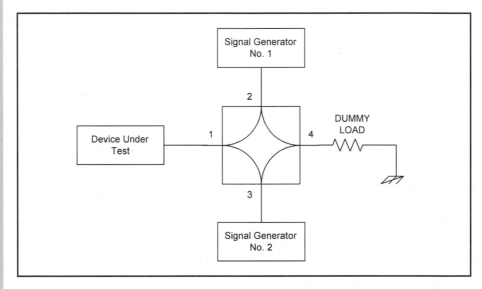

Figure 9-22. Combining two signal sources.

Bidirectional Amplifiers. A number of different applications exist for bidirectional amplifiers; i.e., amplifiers that can handle signals from two opposing directions on a single line. The telecommunications industry, for example, uses such systems to send full-duplex signals over the same lines.

Similarly, cable TV systems that use two-way (e.g., cable MODEM) require two-way amplifiers. *Figure 9-23* shows how the hybrid coupler can be used to make such an amplifier. In some telecommunications text-books the two directions are called East and West, so this amplifier is occasionally called an East-West amplifier. At other times this circuit is called a *repeater*.

In the bidirectional E-W amplifier of *Figure 9-23* amplifier A1 amplifies the signals traveling west-to-east, while A2 amplifies signals traveling east-to-west. In each case, the amplifiers are connected to hybrids HB1 and HB2 via opposite ports, so will not interfere with each other. Otherwise, connecting two amplifiers input-to-output-to-input-to-output is a recipe for disaster...if only a large amount of destructive feedback.

Transmitter/Receiver Isolation. One of the problems that exists when using a transmitter and receiver together on the same antenna is isolating the receiver input from the transmitter input. Even a weak transmitter will burn out the receiver input if its power were allowed to reach the receiver input circuits. One solution is to use one form of transmit/receive (T/R) relay. But that solution relies on an electromechanical device, which adds problems of its own (not the least of which is reliability).

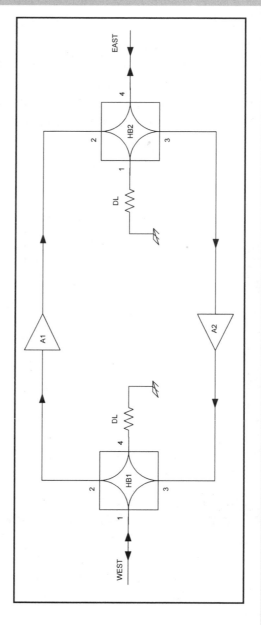

Figure 9-23. Bidirectional "repeater" amplifier.

A solution to the T/R problem using a hybrid is shown in *Figure 9-24*. In this circuit the transmitter output and receiver input are connected to opposite ports of a hybrid device. Thus, the transmitter power does not reach the receiver input.

The antenna is connected to the adjacent port between the transmitter port and the receiver port. Signal from the antenna will flow over the Port-1 to Port-2 path to reach the receiver input. Transmitter power, on the other hand, will enter at Port-3, and is split into two equal portions. Half the power flows to the antenna over the Port-3 to Port-1 path, while half the power flows to a dummy load through the Port-3 to Port-4 path.

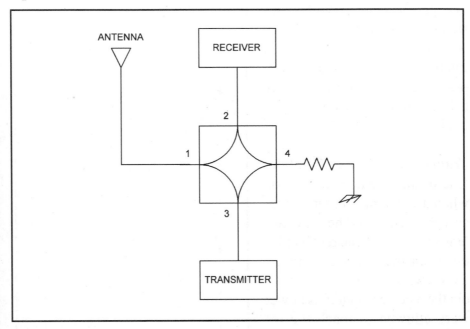

Figure 9-24.
Use of hybrid as a
T/R switch.

There is a problem with this configuration. Because half the power is routed to a dummy load, there is a 3-dB reduction in the power available to the antenna. A solution is shown in *Figure 9-25*. In this configuration a second antenna is connected in place of the dummy load. Depending on the spacing (S), and the phasing, various directivity patterns can be created using two identical antennas.

If the hybrid produces no phase shift of its own, then the relative phase shift of the signals exciting the antennas is determined by the length of the transmission line between the hybrid and that antenna. A 0° phase shift is created when both transmission lines are the same length. Making one transmission line half wavelength longer than the other results in a

180° phase shift. These two relative phase relationships are the basis for two popular configurations of phased array antenna. Consult an antenna book for other options.

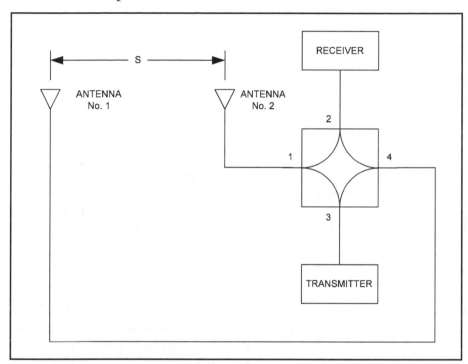

Figure 9-25. Combining two antennas in T/R switch.

Phase-Shifted Hybrids

The hybrids discussed thus far split the power half to each adjacent port, but the signals at those ports are in-phase with each other; that is, there is a zero degree phase shift over the paths from the input to the two output ports. There are, however, two forms of phase shifted hybrids. The form shown in *Figure 9-26A* is a 0°-180° hybrid. The signal over the Port-1 to Port-2 path is not phase shifted (0°), while that between Port-1 and Port-3 is phase shifted 180°. Most transformer-based hybrids are inherently 0°-180° hybrids.

A 0°-90° hybrid is shown in *Figure 9-26B*. This hybrid shows a 90° phase shift over the Port-1/Port-2 path, and a 0° phase shift over the Port-1/Port-3 path. This type of hybrid is also called a *quadrature hybrid*.

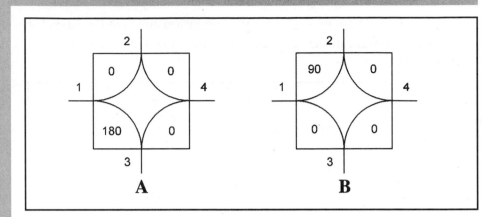

Figure 9-26.
A) 180 degree hybrid;
B) 90 degree hybrid.

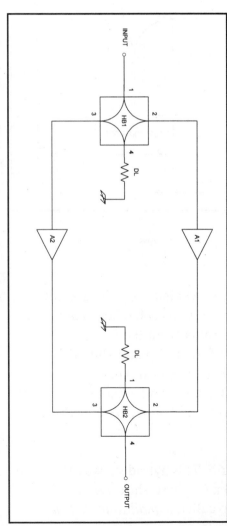

One application for the quadrature hybrid is the balanced amplifier shown in *Figure 9-27*. Two amplifiers, A1 and A2, are used to process the same input signal arriving via hybrid HB1. The signal splits in HB1, so becomes inputs to both A1 and A2. If the input impedances of the amplifiers are not matched to the system impedance, then signal will be reflected from the inputs back towards HB1. The reflected signal from A2 arrives back at the input in-phase (0°), but that reflected from A1 has to pass through the 90° phase shift arm twice, so has a total phase shift of 180°. Thus, the reflections caused by mismatching the amplifier inputs are canceled out.

The output signals of A1 and A2 are combined in hybrid HB2. The phase balance is restored by the fact that the output of A1 passes through the 0° leg of HB2, while

Figure 9-27.
Balanced amplifier.

the output of A2 passes through the 90° leg. Thus, both signals have undergone a 90° phase shift, so are now restored to the in-phase condition.

Use With Receiver Antennas. Examples given above combine a receiver and transmitter on a single antenna or antenna system. It's also possible to use the hybrid for antenna arrays intended for receivers. Antennas spaced some distance X apart will have different patterns and gains depending on the value of X and the relative phase of the currents in the two antennas. One can, therefore, connect the antennas to Ports-2 and 3, and the receiver antenna input to Port-1. A terminating resistor would be used at Port-4. You can use either 0-degree, 90-degree or 180-degree hybrids, depending on the particular antenna system.

Diplexers

The diplexer is a passive RF device that provides frequency selectivity at the output, while looking like a constant resistive impedance at its input terminal. *Figure 9-28* shows a generalization of the diplexer. It consists of a *high-pass filter* and a *low-pass filter* that share a common input line. With appropriate design the diplexer will not exhibit any reactance reflected back to the input terminal (which eliminates the reflections and VSWR problem). Yet, at the same time it will separate the high- and low-frequency components into two separate signal channels.

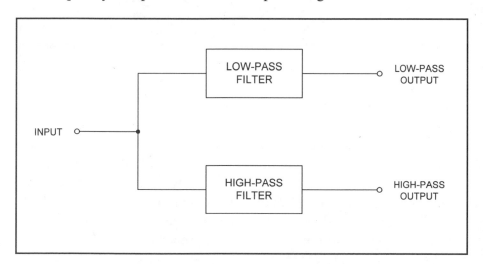

Figure 9-28. Block diagram of a diplexer based on a high-pass and low-pass filter.

Figure 9-29 shows a practical diplexer built with lumped constant inductor (L) and capacitor (C) components. The high-pass filter consists of C1, C2, C3, L1 and L2. The low-pass filter consists of C4, C5, L3, L4 and L5. Both filters share a common ground at the chassis.

The LC component values shown in *Figure 9-29* are for a -3 dB cutoff frequency of 100 MHz for both the high-pass and low-pass filter sections. Note that the corner frequency of the filters is the same for both sections, so there is no portion of the spectrum that is not covered by the filter, except very close to the 100-MHz corner frequency. The circuit's component values can be scaled to other frequencies by using a simple formula:

$$New\,Value \;=\; \frac{Old\,Value}{F_{MHz}} \qquad\qquad (9\text{-}9)$$

Where: F_{MHz} is the desired frequency in megahertz (MHz).

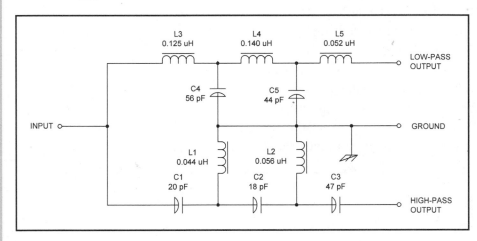

Figure 9-29.
Low-pass/high-pass
diplexer circuit
centered on 100MHz.
Other frequency can
be scaled from these
component values.

Using a Diplexer with a Mixer

One common use for the diplexer circuit is to smooth the impedance excursions seen by a mixer circuit, and to prevent reflected signals from the load back to the mixer from interfering with mixer performance. Reflections can seriously affect the intercept points of the mixer, which means that they can reduce the dynamic range and increase undesirable intermodulation distortion. The diplexer approach of *Figure 9-30* is particularly useful where the mixer and the following circuitry are mismatched at frequencies other than the desired frequency.

Figure 9-30 shows the two cases. In each case, a mixer nonlinearly combines two frequencies, F1 and F2, to produce an output spectrum of mF1 ± nF2, where m and n are integers representing the fundamental and harmonics of F1 and F2. In some cases, we are interested only in the difference frequency, so will want to use the low-pass output (LPO) of the diplexer (*Figure 9-30A*). The high-pass output (HPO) is terminated in a matched load so that signal transmitted through the high-pass filter is fully absorbed in the load.

The exact opposite situation is shown in *Figure 9-30B*. Here we are interested in the sum frequency, so use the HPO port of the diplexer, and terminate the LPO port in a resistive load. In this case, the signal passed through the low-pass filter section will be absorbed by the load.

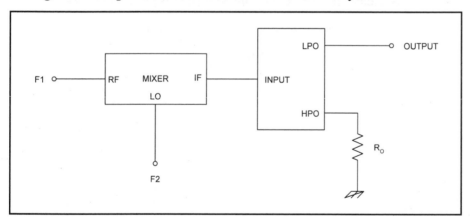

Figure 9-30A.
Use of a diplexer to improve mixer performance for difference IF.

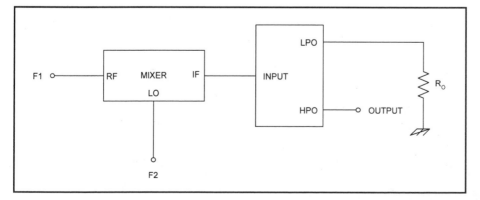

Figure 9-30B.
Use of a diplexer to improve mixer performance for sum IF.

Bandpass Diplexers

Figures 9-31 and *9-32* show two different bandpass diplexer circuits commonly used at the outputs of mixers. These circuits use a bandpass filter approach rather than two separate filters. *Figure 9-31* shows a pp-network approach, while the version in *Figure 9-32* is an L-network. In both cases,

$$Q = \frac{f_o}{BW_{3dB}}$$ (9-10)

and,

$$\omega = 2\pi f_o$$ (9-11)

Where:

f_o is the center frequency of the passband in hertz (Hz)

BW_{3dB} is the desired bandwidth in hertz (Hz)

Q is the relative bandwidth

For the circuit of *Figure 9-31*:

$$L2 = \frac{R_o Q}{\omega}$$ (9-12)

$$L1 = \frac{R_o}{\omega Q}$$ (9-13)

$$C2 = \frac{1}{R_o Q \omega}$$ (9-14)

$$C1 = \frac{Q}{\omega R_o}$$ (9-15)

For the circuit of *Figure 9-32*:

$$L2 \;=\; \frac{R_o \, Q}{\omega} \qquad\qquad (9\text{-}16)$$

$$L1 \;=\; \frac{R_o}{\omega \, Q} \qquad\qquad (9\text{-}17)$$

$$C1 \;=\; \frac{1}{L1 \, \omega^2} \qquad\qquad (9\text{-}18)$$

$$C2 \;=\; \frac{1}{L2 \, \omega^2} \qquad\qquad (9\text{-}19)$$

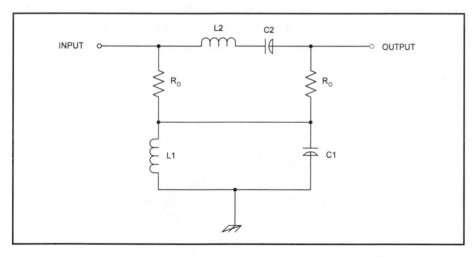

Figure 9-31.
Pi-type bandpass
diplexer.

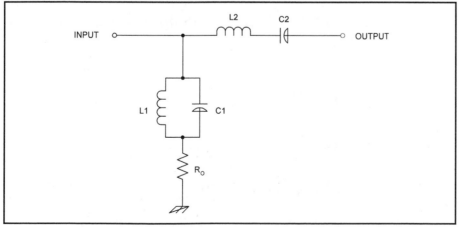

Figure 9-32.
L-type bandpass
diplexer.

Directional Couplers

Directional couplers are devices that will pass signal across one path, while passing a much smaller signal along another path. One of the most common uses of the directional coupler is to sample an RF power signal, either for controlling transmitter output power level or for measurement. An example of the latter use is to connect a digital frequency counter to the low-level port, and the transmitter and antenna to the straight-through (high-power) ports.

A transmission line directional coupler is shown in *Figure 9-33*. The circuit symbol is shown in *Figure 9-33A*. Note that there are three outputs and one input. The IN-OUT path is low-loss, and is the principal path between the signal source and the load. The coupled output is a sample of the forward path, while the isolated port shows very low signal. If the IN and OUT are reversed, then the roles of the coupled and isolated ports also reverse.

An implementation of this circuit using transmission line segments is shown in *Figure 9-33B*. Each transmission line segment (TL1 and TL2) has a characteristic impedance, Z_O, and is quarter wavelength long. The path from Port-1 to Port-2 is the low-loss signal direction. If power flows in this direction, then Port-3 is the coupled port and Port-4 is isolated. If the power flow direction reverses (Port-2 to Port-1) then the respective roles of Port-3 and Port-4 reverse.

For a coupling ratio (Port-3/Port-4) $\leq$ -15 dB, the value of coupling capacitance must be:

$$C_c \quad < \quad \frac{0.18}{\omega Z_o} \text{ farads} \qquad\qquad (9\text{-}20)$$

The coupling ratio is:

$$C.R. \quad = \quad 20\, Log\left(\omega\, C\, Z_o\right) \quad \text{dB} \qquad\qquad (9\text{-}21)$$

The circuit shown in *Figure 9-33C* is an L-C lumped constant version of the transmission lines. This network can be used to replace TL1 and TL2 in *Figure 9-33A*. The values of the components are:

$$L1 = \frac{Z_o}{\omega_o} \qquad (9\text{-}22)$$

$$C1 = \frac{1}{\omega_o \, Z_o} \qquad (9\text{-}23)$$

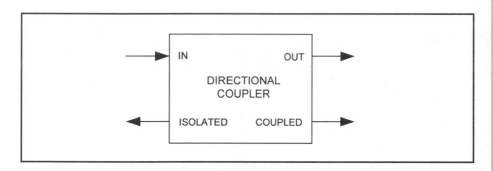

Figure 9-33A.
Directional coupler
symbol.

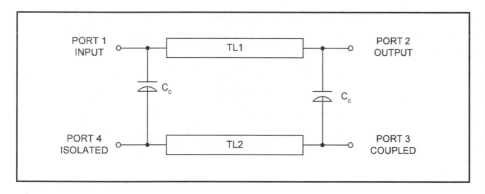

Figure 9-33B.
Transmission line
directional coupler.

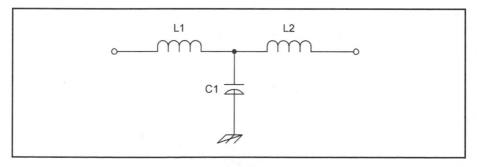

Figure 9-33C.
L-C network to
replace TL1 and
TL2.

Figure 9-34 shows a directional coupler used in a lot of RF power meters and VSWR meters. The transmission lines are implemented as printed-circuit board tracks. It consists of a main transmission line (TL1) between Port-1 and Port-2 (the low-loss path), and a coupled line (TL2) to form the coupled and isolated ports. The coupling capacitance (in pF) is approximated by 9.399X when implemented on G-10 Epoxy fiberglass printed-circuit board.

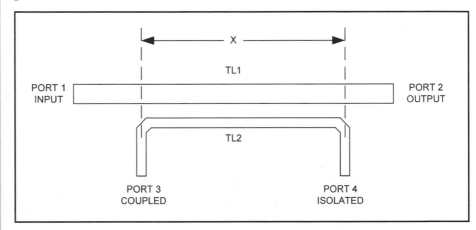

Figure 9-34.
Printed-circuit
directional coupler.

A reflectometer directional coupler is shown in *Figure 9-35A*. This type of directional coupler is at the heart of many commercial VSWR meters and RF power meters used in the HF through low-VHF regions of the spectrum. This circuit is conceptually similar to the previous transmission line, but is designed around a toroid transmission line transformer. It consists of a transformer in which the low-loss path is a single-turn primary winding, and a secondary wound of enameled wire.

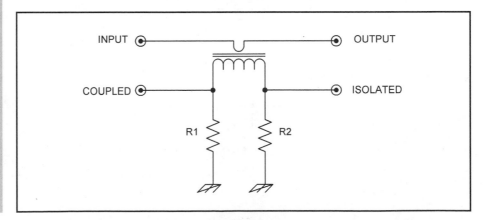

Figure 9-35A.
Toroid core
reflectometer.

Details of the pickup sensor are shown in *Figure 9-35B*. The secondary is wound around the rim of the toroid in the normal manner, occupying not more than 330 degrees of circumference. A rubber or plastic grommet is fitted into the center hole of the toroid core. The single-turn primary is formed by a single conductor passed once through the hole in the center of the grommet. It turns out the 3/16-inch O.D. brass tubing (the kind sold in hobby shops that cater to model builders) will fit through several standard grommet sizes nicely, and will slip-fit over the center conductor of SO-239 coaxial connectors.

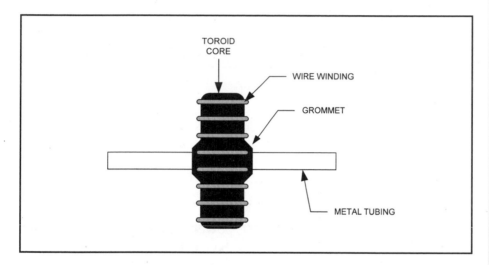

Figure 9-35B.
Toroid transformer
detail.

Another transmission line directional coupler is shown in *Figure 9-36*. Two short lengths of RG-58/U transmission line (≈6 in.) are passed through a pair of toroid coils. Each coil is wound with 8 to 12 turns of wire. Note that the shields of the two transmission line segments are grounded only at one end.

Each combination of transmission line and toroid core form a transformer similar to the previous case. These two transformers are cross-coupled to form the network shown. The XMTR-ANTENNA path is the low-loss path, while (with the signal flow direction shown) the other two coupled ports are for forward and reflected power samples. These samples can be rectified and used to indicate the relative power levels flowing in the forward and reverse directions. Taken together these indications allow us to calculate VSWR.

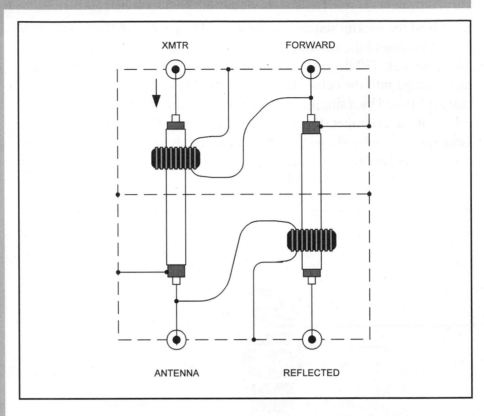

Figure 9-36.
Transmission line
directional coupler.

Directional couplers are used for RF power sampling in measurement and transmitter control. They can also be used in receivers between the mixer or RF amplifier and the antenna input circuit. This arrangement can prevent the flow of LO signal and mixer products back towards the antenna, where they could be radiated and cause electromagnetic interference (EMI) to other devices.

Chapter 10
HF/VHF/UHF LOW-NOISE PREAMPLIFIERS

Chapter 10
HF/VHF/UHF Low-Noise Preamplifiers

RadioScience Observing depends heavily on radio receivers on frequencies from VLF to microwave bands. The performance of receivers can be seriously improved by the use of either a *preselector* or *preamplifier* between the antenna and the receiver. In this chapter we will take a look at small-signal radio frequency (RF) preselectors and preamplifiers. These devices can be used to pre-amplify radio signals from antennas prior to input to a receiver, improve signal-to-noise ratio, and overcome system losses (especially important on radio astronomy receivers in the VHF bands and up).

Most low-priced receivers (and some high-priced ones as well) suffer from performance problems that are a direct result of the tradeoffs the manufacturers have to make in order to produce a low-cost model. In addition, older receivers often suffer the same problems, as do many homebrew radio receiver designs. Chief among these are *sensitivity*, *selectivity* and *image response*.

Sensitivity is a measure of the receiver's ability to pick up weak signals. Part of the cause of poor sensitivity is low gain in the front-end of the radio receiver, although the IF amplifier contributes most of the gain.

Selectivity is a measure of the ability of the receiver to: 1) separate two closely spaced signals; and 2) reject unwanted signals that are not on or very near the desired frequency being tuned. The selectivity provided by a preselector is minimal for very closely spaced signals (that is the job of the IF selectivity in a receiver), but it is used for reducing the effects (e.g., input overloading) of large local signals...so fits the second half of the definition.

Image response affects only superheterodyne receivers (which most are), and is an inappropriate response to a signal that is at a frequency of twice the receiver IF frequency from the frequency that the receiver is tuned to. A superhet receiver converts the signal frequency (RF) to an intermediate frequency (IF) by mixing it with a local oscillator (LO) signal generated inside the receiver. The IF can be either the sum or difference between the LO and RF (i.e., LO+RF or LO-RF), but in most older receivers and nearly all low-cost receivers it is the difference (LO-RF). The problem is that there are always *two* RF frequencies that meet "difference" criteria: LO-IF, and LO+IF. Thus, both LO+IF and LO-IF are equal to the IF frequency. If one is the desired frequency, then the other is the image frequency. If the image frequency gets through the radio's front-end tuning to the mixer, then it will appear in the output as a valid signal.

A cure for all of these problems is a little circuit called the *active preselector*. A preselector can be either active or passive. In both cases, however, the preselector includes a resonant circuit that is tuned to the frequency that the receiver is tuned to. The preselector is connected between the antenna and the receiver antenna input connector (*Figure 10-1*); therefore, it adds a little more selectivity to the front-end of the radio to help discriminate against unwanted signals.

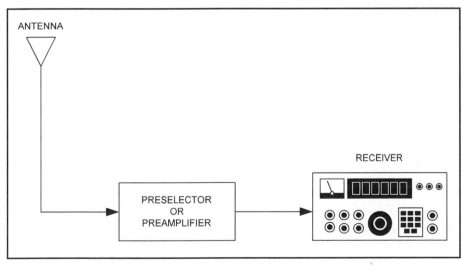

Figure 10-1.
Position of a preamplifier with a receiver.

The difference between the active and passive designs is that the active design contains an RF amplifier stage, while the passive design does not. Thus, the active preselector also deals with the sensitivity problem of the receiver. In this section we will take a look at several active preselector circuits that you can build and adapt to your own needs.

The difference between a preamplifier and the amplifying variety of preselector is that the preselector is tuned to a specific frequency or narrow band of frequencies. The wideband preamplifier amplifies all signals coming into the front-end, with no discrimination, and therein lies an occasional problem.

Another problem with any amplifier ahead of the receiver is that the preamplifier might deteriorate performance, rather than make it better. The amplifier may use up part of the receiver's dynamic range, or it may cause moderate signals that it can handle now to drive the receiver into compression. Along with compression comes intermodulation distortion, possibly harmonic distortion, desensitization and other problems...too many problems to beat. So, make sure you are not going to drive the receiver into poor performance while trying to improve it!

Always use a preamplifier or preselector that has a noise figure that is better than the receiver being served. The Friis equation for noise demonstrates that the noise figure of the system is utterly dominated by the noise figure of the first amplifier. So make sure that the amplifier is a low-noise amplifier (LNA), and has a noise figure a few dB less than the receiver's noise figure.

Note: Preselectors and preamplifiers should be built inside well-shielded metal boxes in order to prevent RF leakage around the device. Select boxes that are either die-cast, or are made of sheet metal and have an overlapping lip. Do not use the lower cost tab-fit sheet metal type box.

Noise and Preselectors/Preamplifiers

The weakest radio signal that you can detect on a receiver is determined mainly by the noise level in the receiver. Some noise arrives from outside sources, while other noise is generated inside the receiver. At the VHF/UHF/microwave range, the internal noise is predominant, so it is common to use a *low-noise preamplifier* ahead of the receiver. The preamplifier will reduce the noise figure for the entire receiver. If you select a commercial ready-built or kit VHF preamplifier, such as the units sold by Hamtronics, Inc., then make sure you specify the low-noise variety for the first amplifier in the system.

Hamtronics, Inc. (NY) 65 Moul Rd., Hilton, NY, 14468-9535; Ph: 716-392-9430

The low-noise amplifier (LNA) should be mounted on the antenna if it is wideband, and at the receiver if it is tunable (NOTE: the term *preselector* only applies to tuned versions, while *preamplifier* can denote either tuned or wideband models). Of course, if your receiver is used only for one frequency, then it may also be mounted at the antenna. The reason for mounting the amplifier right at the antenna is to build up the signal and improve the signal-to-noise ratio (SNR) *prior* to feeding the signal into the transmission line, where losses cause it to weaken somewhat.

A preamplifier or preselector may improve the performance of your receiver, no matter whether you listen to VLF, AM broadcast band, shortwaves or VHF/UHF/microwave bands. The circuits presented in this chapter allow you to "roll your own" and be successful at it.

JFET Preselector

Figure 10-2 shows the most basic form of JFET preselector. This circuit will work into the low-VHF region. This circuit is in the common source configuration, so the input signal is applied to the gate and the output signal is taken from the drain. Source bias is supplied by the voltage drop across resistor R2, and drain load by a series combination of a resistor (R3) and a radio-frequency choke (RFC1). The RFC should be 1000 µH (1 mH) at the AM broadcast band and HF (shortwave), and 100 µH in the low-VHF region (>30 MHz). At VLF frequencies below the broadcast band use 2.5 mH for RFC1, and increase all 0.01 µF capacitors to 0.1 µF. All capacitors are either disk ceramic, or one of the newer dielectric capacitors (*if* rated for VHF service...be careful—not all are!).

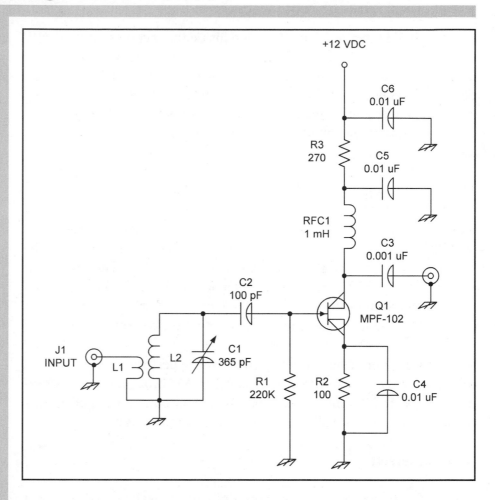

Figure 10-2.
JFET preamplifier
(common source
configuration).

The input circuit is tuned to the RF frequency, but the output circuit is untuned. The reason for the lack of output tuning is that tuning both input and output permits the JFET to oscillate at the RF frequency...and that we don't want. Other possible causes of oscillation include layout, and a *self-resonance* frequency of the RFC that is too near the RF frequency (select another choke).

The input circuit consists of an RF transformer that has a tuned secondary (L2/C1). The variable capacitor (C1) is the tuning control. Although the value shown is the standard 365 pF "AM broadcast variable," any form of variable can be used if the inductor is tailored to it. These components are related by:

$$f = \frac{1}{2\pi\sqrt{LC}} \qquad\qquad (10\text{-}1)$$

Where:

f is the frequency in hertz

L is the inductance in Henrys

C is the capacitance in Farads.

Be sure to convert inductances from microhenrys to Henrys, and pico-farads to Farads. Allow approximately 10 pF to account for stray capacitances, although keep in mind that this number is a guess that may have to be adjusted (it is a function of your layout, among other things). We can also solve Equation 10-1 for either L or C:

$$L = \frac{1}{39.5\, f^2\, C} \qquad\qquad (10\text{-}2)$$

Space does not warrant making a sample calculation, but we can report results for you to check for yourself. In a sample calculation, I wanted to know how much inductance is required to resonate 100 pF (90 pF capacitor plus 10 pF stray) to 10 MHz WWV. The solution, when all numbers are converted to Hertz and Farads, results in 0.00000253 H, or 2.53 µH. Keep in mind that the calculated numbers are close, but are nonetheless approximate...and the circuit may need tweaking on the bench.

The inductor (L1/L2) may be either a variable inductor (as shown) from a distributor such as Digi-Key, or "homebrewed" on a toroidal core. Most people will want to use the T-50-6 (RED) or T-68-6 (RED) toroids for shortwave applications. The number of turns required for the toroid is calculated from:

Digi-Key
P.O. Box 677, Thief
River Falls, MN, 56701;
1-800-344-4539

$$N = 100\sqrt{\frac{L_{\mu H}}{A_L}} \qquad\qquad (10\text{-}3)$$

Where:

$L_{\mu H}$ is in microhenrys

A_L is 49 for T-50-2 (RED), and 57 is for T-68-2 (RED).

Example, a 2.53 µH coil needed for L2 (*Figure 10-2*) wound on a T-50-RED core requires 23 turns. Use #26 or #28 enameled wire for the winding. Make L1 approximately 4-7 turns over the same form as L2.

Be careful when making JFET or MOSFET RF amplifiers in which both input and output are tuned. If the circuit is a common-source circuit—i.e., where the input signal is across the gate and source, and the output signal is between the drain and source—there is the possibility of accidentally turning the circuit into a dandy little oscillator. Sometimes this problem is alleviated by tuning the input and output LC tank circuits to slightly different frequencies. In other cases, it is necessary to neutralize the stage. It is a common practice to make at least one end of the amplifier—usually the output—untuned in order to overcome this problem (although at the cost of some gain).

Figure 10-3 shows two methods for tuning both the input and output circuits of the JFET transistor. In both cases the JFET is wired in the common-gate configuration, so signal is applied to the source and output is taken from the drain. The dotted line indicates that the output and input tuning capacitors are ganged to the same shaft.

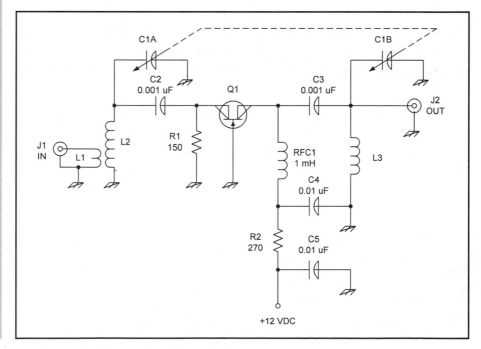

Figure 10-3A.
Common gate
JFET preamplifier,
transformer
coupling, method 1.

The source circuit of the JFET is low-impedance, so some means must be provided to match the circuit to the tuned circuit. In *Figure 10-3A* a tapped inductor is used for L2 (tapped at 1/3 the coil winding), and in *Figure 10-3B* a similar but slightly different configuration is used.

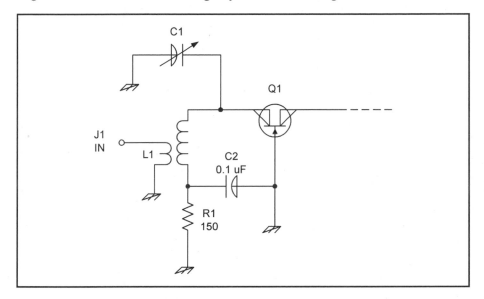

Figure 10-3B. Common gate JFET preamplifier, transformer coupling, method 2.

VHF Receiver Preselector

The circuit in *Figure 10-4* is a VHF preamplifier that uses two JFET devices connected in *cascode*; that is, the input device (Q1) is in common-source and is direct coupled to the common-gate output device (Q2). In order to prevent self-oscillation of the circuit a *neutralization capacitor* (NEUT) is provided. This capacitor is adjusted to keep the circuit from oscillating at any frequency within the band of operation. In general, this circuit is tuned to a single channel by the action of L2/C1 and L3/C2.

MOSFET Preselector

The 40673 dual-gate MOSFET (*Figure 10-5*) used in the following preselector circuit is low-cost and easily available. It is a *dual-gate* MOSFET, so one gate can be used for amplification and the other for DC-based gain control. Signal is applied to gate G1, while gate G2 is

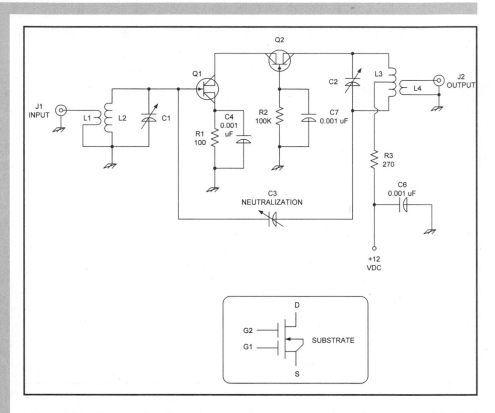

Figure 10-4.
Cascode JFET
preamplifier.

either biased to a fixed positive voltage or connected to a variable DC voltage that serves as a gain control signal. The DC network is similar to that of the previous (JFET) circuits, with the exception that a resistor voltage divider (R3/R4) is needed to bias gate G2.

There are three tuned circuits for this preselector project, so it will produce a large amount of selectivity improvement and image rejection. The gain of the device will also provide additional sensitivity. All three tuning capacitors (C1A, C16 and C1C) are ganged to the same shaft for "single-knob tuning." The trimmer capacitors (C2, C3 and C4) are used to adjust the tracking of the three tuned circuits (i.e., ensure that they are all tuned to the same frequency at any given setting of C1A-C).

The inductors are of the same sort as described above. It is permissible to put L1/L2 and L3 in close proximity to each other, but these should be separated from L4 in order to prevent unwanted oscillation due to feedback arising from coil coupling.

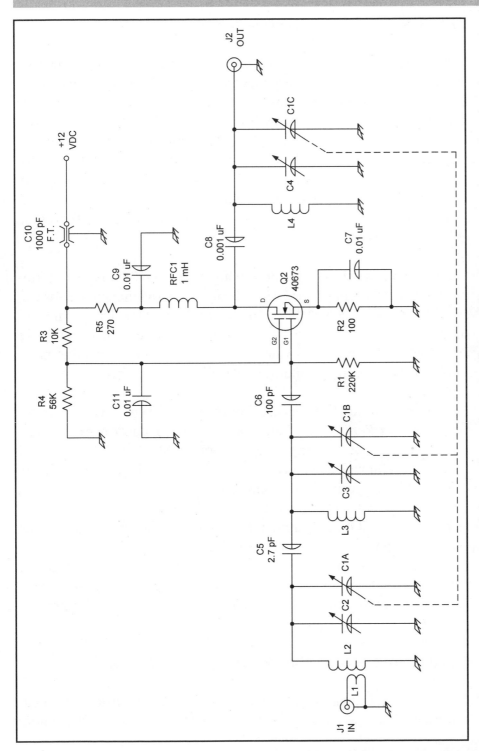

Figure 10-5.
MOSFET
preamplifier.

Voltage Tuned Receiver Preselector

The circuit in *Figure 10-6* is a little different. In addition to using only input tuning (which lessens the potential for oscillation), it also uses *voltage tuning*. The hard-to-find variable capacitors are replaced with *varactor diodes*, also called *voltage variable capacitance diodes*. These PN junction diodes exhibit a capacitance that is a function of the applied reverse bias potential, V_T. Although the original circuit was built and tested for the AM broadcast band (540 kHz to 1700 kHz), it can be changed to any band by correct selection of the inductor values. The designated varactor (NTE-618) offers a capacitance range of 440 pF down to 15 pF over the voltage range 0 to +18 VDC.

The inductors may be either "store-bought" types or wound over toroidal cores. I used a toroid for L1/L2 (forming a fixed inductance for L2) and "store-bought" adjustable inductors for L3 and L4. There is no reason, however, why these same inductors cannot be used for all three uses. Unfortunately, not all values are available in the form that has a low-impedance primary winding to permit antenna coupling.

An aluminum utility box was used for the shielded enclosure, and ordinary *Vector* or *Radio Shack* perf-board is used for constructing the circuit. The RF input and output connectors are SO-239 "UHF" coaxial connectors, although any other type can also be used. Select connectors that match your receiver and antenna system.

In both of the MOSFET circuits the fixed bias network used to place gate G2 at a positive DC potential can be replaced with a variable voltage circuit. The potentiometer in *Figure 10-7* can be used as an RF GAIN control to reduce gain on strong signals, and increase it on weak signals. This feature allows the active preselector to be custom set to prevent overload from strong signals.

Broadband RF Preamplifier for VLF, LF and AM BCB

There are many situations where a broadband RF amplifier is needed. Typical applications include boosting the output of RF signal generators

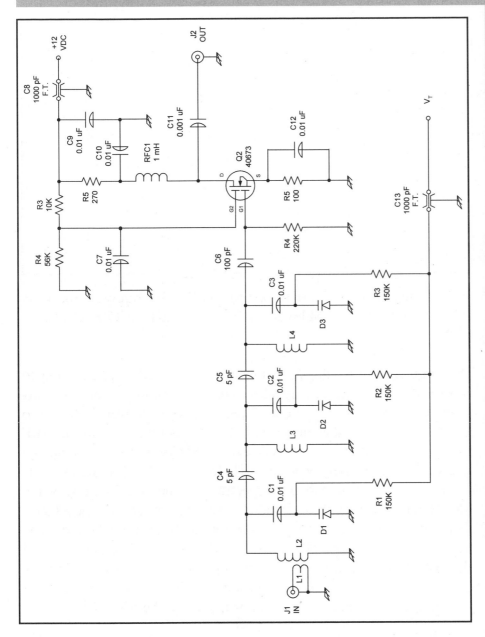

Figure 10-6.
Voltage-tuned
MOSFET
preamplifier.

(which tend to be normally quite low-level), antenna preamplification, loop antenna amplifier, and in the front-ends of receivers. There are a number of different circuits published, including some by me, but one failing that I've noted on most of them is that they often lack response at the low end of the frequency range. Many designs offer -3 dB frequency

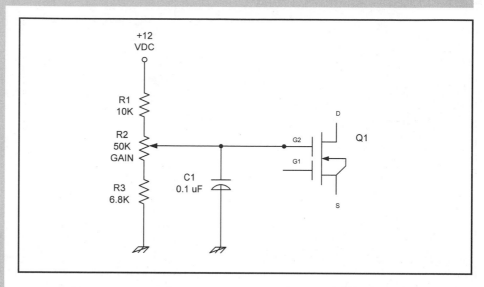

Figure 10-7.
Gain control for
MOSFET RF
amplifier.

response limits of 3 to 30 MHz, or 1 to 30 MHz, but rarely are the VLF, LF or even the entire AM broadcast band (540 kHz to 1700 kHz) covered.

The original need for this amplifier was that I needed an amplifier to boost AM BCB signals. Many otherwise fine communications or entertainment-grade "general coverage" receivers operate from 100 kHz to 30 MHz or so, and that range initially sounds real good to the VLF through AM BCB DXer. But when examined closer it turns out that the receiver lacks sensitivity on the bands below either 2 or 3 MHz, so it fails somewhat in the lower end of the spectrum. While most listening on the AM BCB is to powerful local stations (where receivers with no RF amplifier and a loopstick antenna will work nicely), those who are interested in DXing are not well served. In addition to the receiver, I wanted to boost my signal generator 50-ohm output to make it easier to develop some AM and VLF projects that I am working on, and to provide a preamplifier for a square loop antenna that tunes the AM BCB.

Several requirements were developed for the RF amplifier. First, it had to retain the 50-ohm input and output impedances that are standard in RF systems. Second, it had to have a high dynamic range and third-order intercept point in order to cope with the bone crunching signal levels on the AM BCB. One of the problems of the AM BCB is that those sought-after DX stations tend to be buried under multikilowatt local stations on adjacent channels. That's why high dynamic range, high intercept point and loop antennas tend to be required in these applications. I also wanted the amplifier to cover at least two octaves (4:1 frequency ratio), and in fact achieved a decade (10:1) response (250 kHz to 2,500 kHz).

Furthermore, the amplifier circuit had to be easily modifiable to cover other frequency ranges up to 30 MHz. This last requirement would make the amplifier useful to a large number of readers, as well as extending its usefulness to me.

There are a number of issues to consider when designing an RF amplifier for the front-end of a receiver. The dynamic range and intercept point requirements were mentioned above. Another issue is the amount of distortion products (related to third-order intercept point) that are generated in the amplifier. It does no good to have a high capability on the preamplifier only to overload the receiver with a lot of extraneous RF energy it can't handle...energy that was generated by the preamplifier, not from the stations being received. These considerations point to the use of a *push-pull RF amplifier* design.

Push-Pull RF Amplifiers

The basic concept of a push-pull amplifier is demonstrated in *Figure 10-8*. This type of circuit consists of two identical amplifiers that each process half the input sinewave signal. In the circuit shown, this job is accomplished by using a center-tapped transformer at the input to split the signal, and another at the output to recombine the signals from the two transistors. The transformer splits the signal because its center tap is grounded, and thus serves as the common for the signals applied to the two transistors. Because of normal transformer action, the signal polarity at end "A" will be opposite that at end "B" when the center tap ("CT") is grounded. Thus, the two amplifiers are driven 180 degrees out of phase with each other; one will be turning on while the other is turning off, and vice versa.

The push-pull amplifier circuit is balanced, and as a result it has a very interesting property: even-order harmonics are cancelled in the output, so the amplifier output signal will be cleaner than for a single-ended amplifier using the same active amplifier devices.

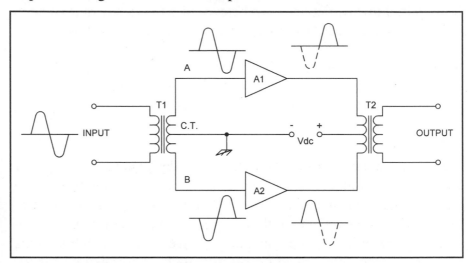

Figure 10-8.
Push-pull broadband amplifier block diagram.

The Push-Pull RF Amplifier

There are two general categories of push-pull RF amplifiers: tuned amplifiers and wideband amplifiers. The tuned amplifier will have the inductance of the input and output transformers resonated to some specific frequency. In some circuits the nontapped winding may be tuned, but in others a configuration such as *Figure 10-9* might be used. In this circuit both halves of the tapped side of the transformer are individually tuned to the desired resonant frequency. Where variable tuning is desired, a split-stator capacitor might be used to supply both capacitances.

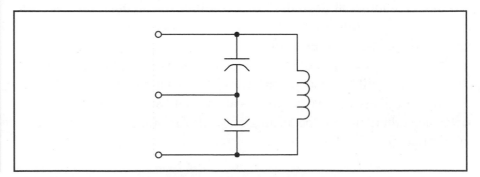

Figure 10-9.
Tuned coupling transformer.

The broadband category of circuit is shown in *Figure 10-10A*. In this type of circuit a special transformer is usually needed. The transformer must be a broadband RF transformer, which means that it must be wound on a suitable core such that the windings are bifilar or trifilar. The particular transformer in *Figure 10-10A* has three windings, of which one is much smaller than the others. These must be trifilar-wound for part of the way, and bifilar the rest of the way. This means that all three windings are kept parallel until no more turns are required of the coupling link, and then the remaining two windings are kept parallel until they are completed. *Figure 10-10B* shows an example for the case where the core of the transformer is a ferrite or powdered-iron *toroid*.

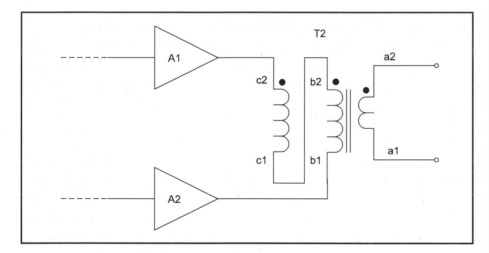

Figure 10-10A.
Untuned
(broadband)
coupling
transformer.

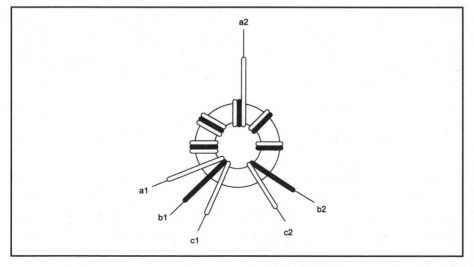

Figure 10-10B.
Winding the
transformer on a
toroid.

Actual Circuit Details

The actual RF circuit is shown in *Figure 10-11*. The active amplifier devices are junction field-effect transistors (JFET) intended for service from DC to VHF. The device selected can be the ever-popular MPF-102, or its replacement equivalent from the SK, ECG or NTE lines of devices. Also useful is the 2N4416 device. The particular device that I used was the NTE-451 JFET transistor. This device offers a transconductance of 4,000 microsiemens[*], a drain current of 4 to 10 mA, and a power dissipation of 310 mW, with a noise figure of 4 dB maximum.

*Note units:
1 μSiemen = 1 μMho

The JFET devices are connected to a pair of similar transformers, T1 and T2. The source bias resistor (R1) for the JFETs, and its associated bypass capacitor (C1), are connected to the center tap on the secondary winding of transformer T1. Similarly, the +9 VDC power supply voltage is applied through a limiting resistor (R2) to the center tap on the primary of transformer T2.

Take special note of those two transformers. These transformers are known generally as wideband transmission line transformers, and can be wound on either toroid or binocular ferrite or powdered-iron cores. For the project at hand, because of the low frequencies involved, I selected a type BN-43-202 binocular core. The type 43 material used in this core is a good selection for the frequency range involved. The core can be obtained from *Ocean State Electronics* . There are three windings on each transformer. In each case, the "B" and "C" windings are 12 turns of #30 AWG enameled wire wound in a bifilar manner. The coupling link in each is winding "A." The "A" winding on transformer T1 consists of four turns of #36 AWG enameled wire, while on T2 it consists of two turns of the same wire. The reason for the difference is that the number of turns in each is determined by the impedance matching job it must do (T1 has a 1:9 primary/secondary ratio, while T2 has a 36:1 primary/secondary ratio). Neither the source nor drain impedances of this circuit are 50 ohms (the system impedance), so there must be an impedance transformation function. If the two amplifiers in the circuit were of the sort that had 50-ohm input and output impedances, such as the *Mini-*

Ocean State
Electronics
P.O. Box 1458,
6 Industrial Drive,
Westerly, RI, 02891;
Phones:
401-596-3080 voice,
401-596-3590 fax, or
1-800-866-6626 for
orders only

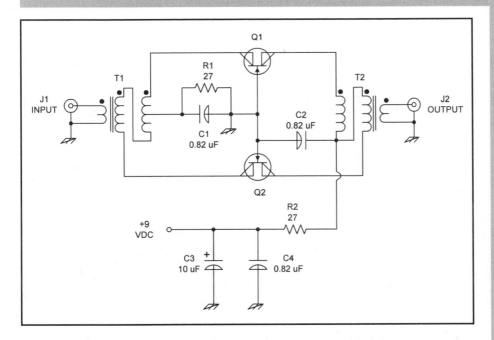

Figure 10-11.
JFET push-pull
broadband
amplifier.

Circuits MAR-1 through MAR-8 devices, then winding "A" in both transformers would be identical to windings "B" and "C." In that case, the impedance ratio of the transformers would be 1:1:1.

The detail for transformers T1 and T2 is shown in *Figure 10-12*. I elected to build a header of printed-circuit perforated board for this part; the board holes are on 0.100 inch centers. The PC type of perf-board has a square or circular printed-circuit soldering pad at each hole. A section of perf-board was cut with a matrix of five holes by nine holes. *Vector Electronics* push terminals are inserted from the unprinted side, and then soldered into place. These terminals serve as anchors for the wires that will form the windings of the transformer. Two terminals are placed at one end of the header, and three at the opposite end.

The coupling winding is connected to pins 1 and 2 of the header, and is wound first on each transformer. Strip the insulation from a length of #36 AWG enameled wire for about 1/4 inch from one end. This can be done by scraping with a scalpel-blade *X-acto* knife, or by burning with the tip of a soldering pencil. Ensure that the exposed end is tinned with solder, and then wrap it around terminal #1 of the header. Pass the wire

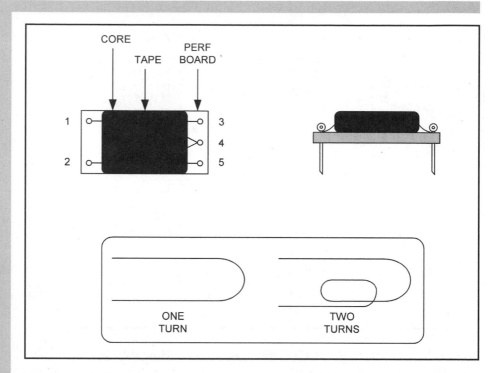

Figure 10-12.
Winding the coupling
transformers using a
bazooka core and a
perf-board or printed-
circuit header.

through the first hole of the binocular core, across the barrier between
the two holes, and then through the second hole. This "U" shaped turn
counts as one turn. To make transformer T1, pass the wire through both
sets of holes three more times (to make four turns). The wire should be
back at the same end of the header as it started. Cut the wire to allow a
short length to connect to pin #2. Clean the insulation off this free end,
tin the exposed portion and then wrap it around pin #2 and solder. The
primary of T1 is now completed.

The two secondary windings are wound together in the bifilar manner,
and consist of 12 turns each of #30 AWG enameled wire. The best approach seems to be twisting the two wires together. I use an electric drill
to accomplish this job. Two pieces of wire, each 30 inches long, are
joined together and chucked up in an electric drill. The other ends of the
wire are joined together and anchored in a bench vise, or some other
holding mechanism. I then back off, holding the drill in one hand, until
the wire is nearly taut. Turning on the drill causes the two wires to twist
together. Keep twisting them until you obtain a pitch of about eight to
twelve twists per inch.

It is VERY IMPORTANT to use a drill that has a variable speed control so that the drill chuck can be made to turn very slowly. It is also VERY IMPORTANT that you follow certain safety rules, especially regarding your eyesight, when making twisted pairs of wire. Be absolutely sure to wear either safety glasses or goggles while doing this operation. If the wire breaks—and that is a common problem—then it will whip around as the drill chuck turns. While #36 wire doesn't seem to be very substantial, at high speed it can severely injure an eye.

To start the secondary windings, scrape all of the insulation off both wires at one end of the twisted pair, and tin the exposed ends with solder. Solder one of these wires to pin #3 of the header, and the other to pin #4. Pass the wire through the hole of the core closest to pin #3, around the barrier, and then through the second hole, returning to the same end of the header as where you started. That constitutes one turn. Now do it 11 more times until all 12 turns are wound. When the 12 turns are completed, cut the twisted pair wires off to leave about 1/2 inch free. Scrape and tin the ends of these wires.

Connecting the free ends of the twisted wire is easy, but you will need an ohmmeter or continuity tester to see which wire goes where. Identify the end that is connected at its other end to pin #3 of the header, and connect this wire to pin #4. The remaining wire should be the one that was connected at its other end to pin #4 earlier; this wire should be connected to pin #5 of the header.

Transformer T2 is made in the identical manner as transformer T1, but with only two turns on the coupling winding rather than four. In this case, the coupling winding is the secondary, while the other two form two halves of the primary. Wind the two-turn secondary first, as was done with the four-turn primary on T1.

The amplifier can be built on the same sort of perforated board as was used to make the headers for the transformers. Indeed, the headers and the board can be cut from the same stock. The size of the board will depend somewhat on the exact box you select to mount it in. For my purposes, the box was a *Hammond* 3" x 5.5" x 1.5" cabinet. Allowing room for the 9 VDC battery at one end, and the input/output jacks and power switch at the other, left me with 2.5" x 3.5" of available space in which to build the circuit (*Figure 10-13*).

I built the circuit from the output end backwards toward the input, so transformer T2 was mounted first with pins 1 and 2 towards the end of the perf-board. Next, the two JFET devices were mounted, and then T1 was soldered into place. After that, the two resistors and capacitors were added to the circuit. Connecting the elements together, and providing push terminals for the input, output, DC power supply ground and the +9 VDC finished the board.

Because the input and output jacks are so close together, and because the DC power wire from the battery to the switch had to run the length of the box, I decided to use a shield partition to keep the input and output separated. This partition was made from 1" brass stock. This material can be purchased at almost any hobby shop that caters to model builders. The RG-174/U coaxial cable between the input jack on the front panel and the input terminals on the perf-board run on the outside of the shield partition.

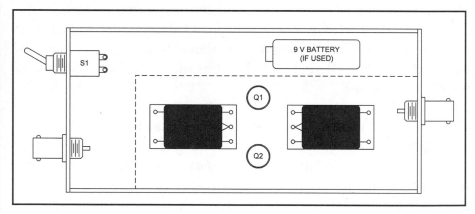

Figure 10-13A. Completed preamplifier.

Variations on the Theme

Three variations on the circuit extend the usefulness for many different readers. First, there are those who want to use the amplifier at the output of a loop antenna that is remote mounted. It isn't easy to go up to the roof or attic to turn on the amplifier any time you wish to use the loop antenna. Therefore, it is better to install the 9 VDC power source at the receiver end, and pass the DC power up the coaxial cable to the amplifier and antenna. This method is shown in *Figure 10-13B*. At the receiver end, RF is isolated from the DC power source by a 10 mH RF choke (RFC2), while the DC is kept from affecting the receiver input (which could short it to ground!) by using a blocking capacitor (C4). All of these components should be mounted inside a shielded box. At the amplifier end, lift the grounded side of the T2 secondary and connect it to RFC2, which is then connected to the +9 VDC terminal on the perfboard. A decoupling capacitor (C3) serves to keep the "cold" end of the T2 secondary at ground potential for RF, while keeping it isolated from ground for DC.

A second variation is to build the amplifier for shortwave bands. This is accomplished easily enough. First, reduce all capacitors to 0.1 μF. Second, build the transformers (T1 and T2) on a toroid core rather than the binocular core. In the original design, a type TF-37-43 ferrite core was used with the same 12:12:2 and 12:12:4 turns scheme as used above.

Alternatively, select a powdered-iron core such as T-50-2 (RED) or T-50-6 (YEL). I suspect that about 20 turns will be needed for the large windings, four turns for the "A" winding on T2 and seven turns for the "A" winding on T1. You can experiment with various cores and turns counts to optimize for the specific section of the shortwave spectrum that you wish to cover.

The third variation is to make the amplifier operate on a much lower frequency; e.g., well down into the VLF region. The principal changes needed are in the cores used for transformers T1 and T2, the number of turns of wire needed, and the capacitors needed. The type 43 core will work down to 10 kHz, or so, but requires a lot more turns to work efficiently in that region. The type 73 material, which is found in the BN-73-202 core, will provide an A_L value of 8,500, as opposed to 2,890 for the BN-43-202 device used in this chapter. Doubling the number of turns in each winding is a good starting point for amplifiers below 200 kHz. The type 73 core works down to 1 kHz, so with a reasonable number of turns should work in the 20 to 100 kHz range as well.

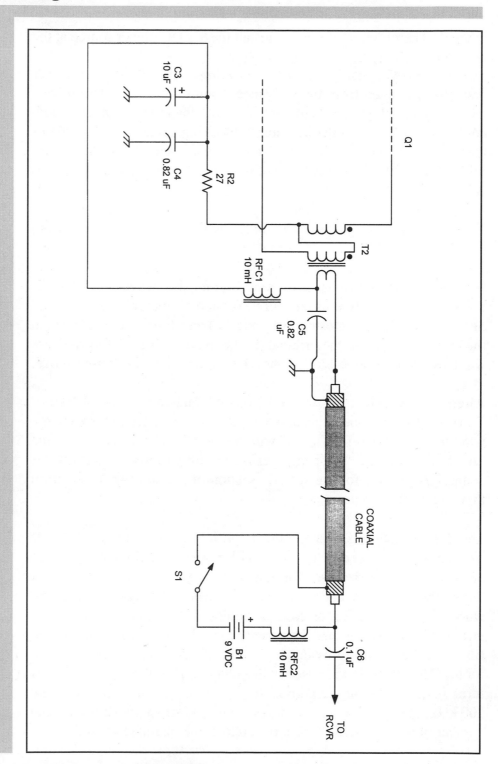

Figure 10-13B.
Powering a remote
preamplifier.

Broadband RF Amplifier (50-Ohm Input and Output)

This project is a highly useful RF amplifier that can be used in a variety of ways. It can be used as a preamplifier for receivers operating in the 3-to-30 MHz shortwave band. It can also be used as a postamplifier following filters, mixers and other devices that have an attenuation factor. It is common, for example, to find that mixers and crystal filters have a signal loss of 5 to 8 dB (this is called "insertion loss"). An amplifier following these devices will overcome that loss. The amplifier can also be used to boost the output level of signal generator and oscillator circuits. In this service it can be used either alone, in its own shielded container, or as part of another circuit containing an oscillator circuit.

The circuit is shown in *Figure 10-14*. The transistor (Q1) is a 2N5179 broadband RF transistor. It can be replaced by the NTE-316 or ECG-316 devices, if the original is not available to you. The NTE and ECG devices are intended for service and maintenance replacement applications, so tend to be found in local electronics parts distributors.

There are two main features to this amplifier: the degenerative feedback in the emitter circuit, and the feedback from collector to base. Degenerative, or negative, feedback is used in amplifiers to reduce distortion (i.e., make it more linear) and to stabilize the amplifier.

One of the negative feedback mechanisms of this amplifier is seen in the emitter. The emitter resistance consists of two resistors—R5 is 10 ohms and R6 is 100 ohms. In most amplifier circuits the emitter resistor is bypassed by a capacitor to set the emitter of the transistor at ground potential for RF signals, while keeping it at the DC level set by the resistance. In normal situations, the reactance of the capacitor should be not more than one-tenth the resistance of the emitter resistor. The 10-ohm portion of the total resistance is left unbypassed, forming a small amount of negative feedback.

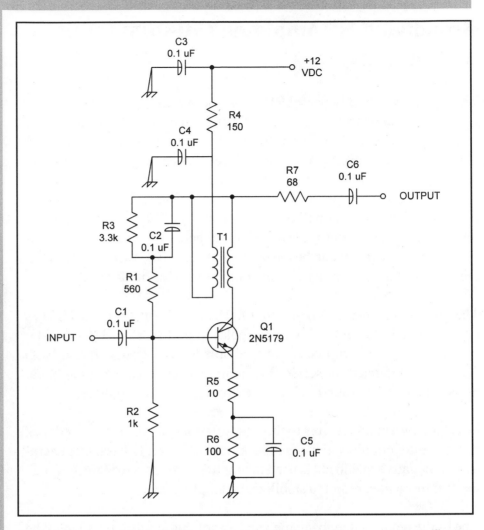

Figure 10-14.
Feedback NPN
transistor
preamplifier.

The collector-to-base feedback is accomplished by two means: first, a resistor-capacitor network (R1/R3/C2) is used; second, a 1:1 broadband RF transformer (T1) is used. This transformer can be homemade. Wind 15 bifilar turns of #26 enameled wire on a toroidal core such as the T-50-2 (RED) or T-50-6 (YEL); smaller cores can also be used.

The circuit can be built on perforated wire-board that has a grid of holes on 0.100-inch centers. You can use a homebrew RF transformer made on a small toroidal core. Use the size 37 core, with #36 enameled wire. As in the previous case, make the two windings bifilar.

Broadband or Tuned RF/IF Amplifier Using the MC-1350P

The MC-1350P is a variant of the MC-1590 device, but unlike the 1590, it is available in the popular, and easy-to-use 8-pin mini-DIP package. It has gain sufficient to make a 30 dB amplifier, although it is a bit finicky and tends to oscillate if the circuit is not built correctly. Layout, in other words, can be a very critical factor because of the gain.

Readers who cannot find the MC-1350P are advised to seek the NTE-746 or ECG-746. These devices are MC-1350Ps, but are sold in the service and maintenance replacement lines, and are usually available locally.

Figure 10-15 shows the basic circuit for the MC-1350P amplifier. The signal is applied to the -IN input, pin #4, while the +IN input is decoupled to ground with a 0.1 µF capacitor. All capacitors in this circuit, except C6 and C7, should be disk ceramic, or one of the newer dielectrics that

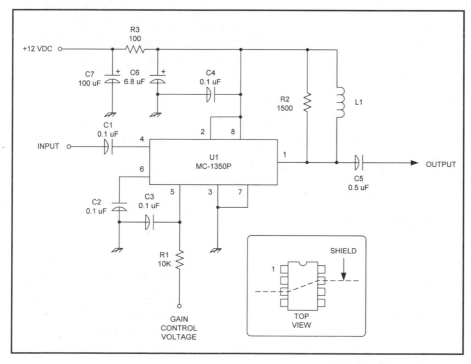

Figure 10-15.
MC-1350P
preamplifier

are competent at RF frequencies to 30 MHz. A capacitor in series with the input terminal, C1, is used to prevent DC riding on the signal from affecting the internal circuitry of the MC-1350P.

The output circuitry is connected to pin #1 of the MC-1350P. Because this circuit is broadband, the output impedance load is a radio-frequency choke (L1). For most HF application, L1 can be a 1 mH choke, although for the lower end of the shortwave region, the medium-wave band and the AM broadcast band, use a 2.5 mH choke. The same circuit can be used for 455 kHz IF amplifier service if the coil (L1) is made 10 mH.

Pin #5 of the MC-1350P device is used for gain control. This terminal needs to see a voltage of +5 to +9 volts, with the maximum gain being found at the +5-volt end of the range (this is opposite what is seen in other chips). The gain control pin is bypassed for RF signals.

The DC power supply is connected to pins 8 and 2 simultaneously. These pins are decoupled to ground for RF by capacitor C4. The ground for both signals and DC power are at pins 3 and 7. The V+ is isolated some-what by a 100-ohm resistor (R3) in series with the DC power supply line. The V+ line is decoupled on either side of this resistor by electro-lytic capacitors. C6 should be a 4.7 to 10 μF tantalum unit, while C7 is a 68 μF (or greater) tantalum or aluminum electrolytic capacitor.

A partial circuit with an alternate output circuit is shown in *Figure 10-16*. This circuit is tuned, rather than broadband, so might be used for IF amplification or RF amplification at specific frequencies. Capacitor C8 is connected in parallel with the inductance of L1, tuning L1 to a specific frequency. In order to keep the circuit from oscillating, the reso-nant tank circuit is "de-Qed" by connecting a 2.2 kohm resistor in paral-lel with the tank circuit. Although considered optional in *Figure 10-16*, it is not optional in this circuit if you want to prevent oscillation.

The MC-1350P device has a disgusting tendency to oscillate at higher gains. One perf-board version of the *Figure 10-15* circuit that I built would not produce more than 16 dB without breaking into oscillation. One tactic to present the oscillation is to use a shield between the input

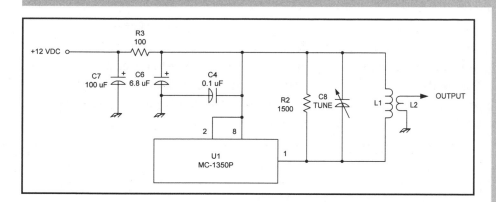

Figure 10-16.
Alternate MC-1350P
output circuit.

and output of the MC-1350P. There are extra holes on the printed-circuit pattern in order to anchor a shield. The shield should be made of copper or brass stock sheet metal, such as the type that can be bought at hobby shops. Cut a small notch along one edge of a piece of 1" stock. The notch should be just large enough to fit over the MC-1350P without shorting out. The location of the shield is shown by the dotted line in *Figure 10-15*. Note that it is bent a little bit in order to fit from the two ground pins on the MC-1350P (i.e., pins 2 and 7).

VLF Preamplifier

The VLF bands run from 5 or 10 kHz to 500 kHz, or just about everything below the AM broadcast band. The frequencies above about 300 kHz can be accommodated by circuitry not unlike 455 kHz IF amplifiers. But as frequency decreases it becomes more of a problem to build a good preamplifier. This project is a preamplifier designed for use from 5 kHz to 100 kHz. This band contains a lot of Navy communications, as well as the most accurate time and frequency station operated by the National Institute of Science and Technology (NIST), WWVB. The operating frequency of WWVB is a very accurate 60 kHz, and is used as a frequency standard in many situations. The signal of WWVB is also used to update electronic clocks.

The circuit of *Figure 10-17* is similar to *Figure 10-15*, except that a large-value RF choke (L1) is used across the -IN and +IN terminals. Both of these chokes are 120 mH size 10 Toko units. If oscillation occurs, then select a different value (82 mH or 100 mH) for one of these chokes. The problem is that the chokes have a capacitance between windings, and that capacitance can resonate the choke to a frequency (this is the "self-resonance" factor). By using different values of choke in input and output circuits, we move their self-resonant frequencies away from each other, reducing the chance of oscillation.

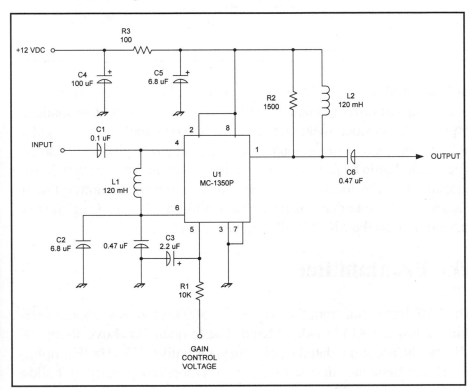

Figure 10-17.
Circuit for MC-1350P
preamplifier
(alternate version).

Chapter 11
RADIO TELEMETRY ON A BUDGET

Chapter 11
Radio Telemetry on a Budget

Radio telemetry is the means for transmitting data from the point of collection to a point where it can be recorded, analyzed and stored. There are a number of different uses for radio telemetry. One use is seen in some hospital coronary care units. When the patient progresses the physician might decide to get them ambulatory, but because there is still a risk, the patient must remain in a defined area and be monitored by radio. A special micro-power radio transmitter collects the electrocardiograph data and transmits it to a radio receiver at or near the nurses' station.

Another case might be where you want to record one or more physical parameters at a remote location. For example, you might want to record the ultraviolet content of the sunlight (see Chapter 8), the temperature, or seismic activity. You could use a digital data logger or a strip chart recorder with an endless paper supply, and come to the site periodically to collect data. But do you really want to do that? And if there are numerous sites the chore might be difficult or even impossibly expensive.

Another use is to track animals in the wild. A beastie is captured and fitted with a radio transmitter. In this case, the data sought might be the location of the animal, so the transmitter might be a simple beacon device, and not transmit any data. On the other hand, it might be encoded to let you know which of several animals is being received (this is especially important if the same frequency is used for several animals).

Figure 11-1 shows a notional radio telemetry system. At the remote site some sort of sensor is used to convert the physical parameter being collected into an analog current or voltage. In some cases, the analog signal will be applied to an analog-to-digital converter (A/D), before being applied to the modulator and transmitter. Some sensors produce digital

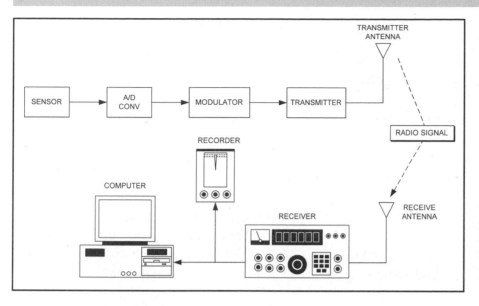

Figure 11-1.
Telemetry set-up.

output, but those include an internal A/D converter. In other cases, the "A/D" converter is replaced by a voltage-to-frequency converter (V/F) that will produce an audio frequency proportional to the applied voltage. Whichever is used, the composite signal will be applied to a transmitter, which produces a radio frequency (RF) signal, and sends it to the antenna. The antenna radiates it to a remote receiver site.

The receiver is tuned to the transmitter frequency. It amplifies and demodulates the signal, to recover the original data signal. This signal might either be displayed on a strip chart or other form of paper recorder, or input to a computer (increasingly the case today).

There are a number of low-cost alternatives to the radio transmitter end of the circuit. One method is to use those little handheld child's walkie-talkies. The audio signal derived from the analog data signal is connected internally to the microphone input line (note: this might also be the loud-speaker in most models). The push-to-talk switch is then wired into the "transmit" position permanently.

I received a letter from an amateur rocketry enthusiast who used the printed-circuit board recovered from a $10 49-MHz walkie-talkie as the telemetry transmitter aboard his rockets. He had a barometric pressure sensor to give him an idea of the altitude. The sensor was used to modulate a V/F converter that then was applied to the transmitter's audio input.

Another rocketry application was seen in a science fair that I judged. The kid used a similar 49-MHz walkie-talkie to monitor the spin rate of his "bird." He used a phototransistor and operational amplifier to modulate a V/F converter. By monitoring the frequency translations as the rocket rotated on its axis (facing the Sun once on each rotation), he was able to record the spinning of the rocket.

One problem might turn up, however. It seems that the audio frequency response of the walkie-talkie is probably limited to the communications equipment standard of 300 to 3,000 Hz (the speech spectrum). If the required audio range exceeds this range, then you might lose data.

A number of frequencies are set aside for this type of application. In the 49-MHz band there is a 100-kHz range that permits up to 100 milliwatts of RF power, relatively large antennas, and speech modulation. Other frequencies in North America include:

> 303.825 MHz
>
> 418 MHz
>
> 433.920 MHz
>
> 916.5 MHz

In other parts of the world:

FREQUENCY	BANDWIDTH	POWER	COMMENTS
173 MHz	150 kHz	10 mW	UK
418 MHz	200 kHz	0.25 mW	UK
433 MHz	1700 kHz	10 mW	Pan-Europe
458 MHz	500 kHz	500 mW	UK
868 MHz	2000 kHz	25/500 mW	Pan-Europe
1.3 GHz	20 MHz	100 mW	UK (video only)
2.4 GHz	30 MHz	100 mW	Worldwide (spread spectrum)

The amount of power required for any given application depends on a lot of factors, but there are some rules of thumb for reasonably flat terrain (i.e., not big hills in the way).

1. 50 meters: 1 mW and a simple antenna

2. 500 meters: 10 mW with a good antenna in a good location.

3. 5,000 meters: 100 mW with a good antenna raised at least 10 meters above mean ground level.

You will find that a 1000 bit/second data rate is easily obtained with audio grade transmitters, while for 10 kilobits/second a narrowband 25-kHz channel is needed. Go to 20 kilobits/second and you will need a wideband radio system. At 100 kilobits/second a professional system is probably needed.

A number of firms offer small printed-circuit board transmitters, receivers, or transmitter/receiver combinations that can be used for telemetry applications. These products are usually designed to be used as if they were integrated circuits (i.e., after the manner of *Basic Stamps* and *PicoStix* processors).

Antennas

A number of different types of antennas can be used for the telemetry transmitter and receiver. In this section we will look at several forms: VHF/UHF Yagi, Normal Mode Helix, Axial Mode Helix and the "Stick."

VHF/UHF Yagi

The Yagi antenna is one of several forms of unidirectional beam antenna. *Figure 11-2* shows the basic form of Yagi (more properly called "Yagi-Uda array" after the two Japanese inventors). The transmission line is connected to the driven element (D.E.), which is essentially a half wavelength dipole. The reflector (REFL) is positioned behind the driven element, opposite the direction of maximum transmission and reception. Directors (DIRx) are positioned in front of the driven element. Spacing and element length are relatively critical, although the antenna will work to one extent or the other if variations exist.

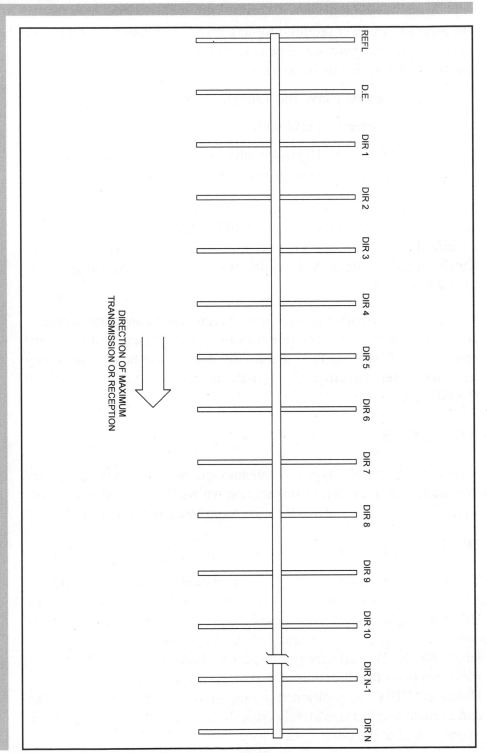

Figure 11-2.
Yagi antenna.

Table 11-1 shows the positions and lengths of the elements for up to a 30-element Yagi centered on 432 MHz. The lengths are taken in millimeters (mm) along the boom that supports the elements. The boom can be made of either wood or aluminum. All elements other than the driven element can be mounted directly to the boom, even if it is made of metal. The driven element, however, must be insulated from the boom.

ELEMENT	POSITION (mm)	LENGTH (mm)
REF	0	340
D.E.	103	335
DIR 1	145	315
DIR 2	223	306
DIR 3	332	300
DIR 4	467	296
DIR 5	622	291
DIR 6	798	290
DIR 7	991	287
DIR 8	1195	284
DIR 9	1415	282
DIR 10	1643	281
DIR 11	1880	279
DIR 12	2120	278
DIR 13	2374	277
DIR 14	2630	276
DIR 15	2890	275
DIR 16	3155	273
DIR 17	3423	272
DIR 18	3693	271
DIR 19	3966	270
DIR 20	4240	268
DIR 21	4520	267
DIR 22	4798	267
DIR 23	5080	266
DIR 24	5362	266
DIR 25	5644	265
DIR 26	5925	265
DIR 27	6210	264
DIR 28	6495	264
DIR 29	6780	263
DIR 30	7066	263

Table 11-1.
Yagi element
lengths.

The reference system puts the reflector (REFL) at position 0-mm, and then gives the positions relative to the reflector end in millimeters.

But suppose you don't want to use 432 MHz (an amateur radio frequency). But rather, you want to use 418 MHz (a telemetry frequency. The antenna can be scaled using a simple fraction, providing that all of the elements and spacings are scaled the same amount. The general equation is:

$$N.D. = \frac{432}{F_{NEW}} \times O.D. \qquad (11\text{-}1)$$

Where:

N.D. is the new dimension (spacing or length)

O.D. is the old dimension from *Table 11-1*

F_{NEW} is the new frequency in MHz

Thus, in our 418 MHz example, the new length of the reflector (340 mm is old length) is:

$$N.D. = \frac{432}{418} \times 340 - mm$$

The nominal feedpoint impedance of a half wavelength dipole like the driven element is around 73 ohms, so it makes a good match for 75-ohm coaxial cable. However, when additional elements are added (for example, reflectors and directors) the impedance drops and is no longer a good match. One method for effecting a match is shown in *Figure 11-3*. It uses a 4:1 VHF/UHF BALUN transformer to make the match. The coax side ("Z = 1" in "4:1") is connected to the coaxial cable from the receiver. The other side is connected to points equally distant from the center point on the driven element. The exact distance is found by measuring the VSWR, while moving the connection points until as low a VSWR as possible is obtained (ideally 1:1, but anything below 1.5:1 will work nicely for most purposes).

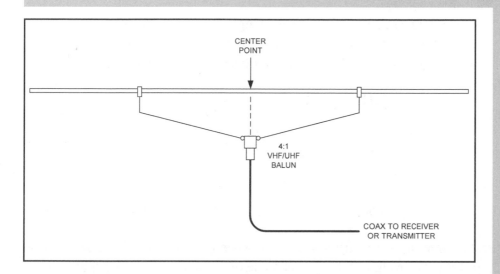

CENTER
POINT

4:1
VHF/UHF
BALUN

COAX TO RECEIVER
OR TRANSMITTER

Figure 11-3.
Simple feed of
the Yagi.

The BALUN can be a television receiver type if this antenna is used for receivers or very, very low-power transmitters. For higher-power transmitters, select an appropriate BALUN from the amateur radio market.

The dimensions for this Yagi assume that the elements are made from 3/16-inch tubing. Other sizes can be used as well, but the dimensions might change a little bit for optimum performance. However, reasonably decent performance can be obtained over a wide range of new frequencies with 1/4-inch to 3/8-inch tubing. The frequency should be within 30 percent or so of 432 MHz, but at least decent performance can be realized over about 50 to 1,500 MHz.

VHF/UHF Ground Plane Antenna

Figure 11-4 shows a vertically polarized ground plane antenna for VHF and UHF bands. This type of antenna produces a vertically polarized omnidirectional radiation and reception pattern. It can therefore be used for a transmitting site where the receiver might be in any direction, or the opposite; i.e., a receiving site where one or more transmitters might be at various azimuthal angles.

The antenna of *Figure 11-4* uses a single vertical radiator element that is quarter wavelength long, and four drooping radials, which are also quarter wavelength long. The lengths of these elements can be found from:

$$L = \frac{7041}{F_{MHZ}} \quad cm \qquad (11\text{-}2)$$

Where:

L is the length of a radiator or radial in centimeters (cm)

F_{MHz} is the design frequency in megahertz (MHz)

The antenna is constructed on a chassis mountable SO-239 coaxial connector. The radiator element is made of copper or brass rod that is sized to slip over the solder connector for the center conductor on the SO-239. You will find lengths of brass tubing of suitable dimensions at hobby shops (the sort that cater to model builders) and craft shops.

The four radials are made from either very stiff solid copper wire (#8 to #12), or brass rods (which can be obtained in the same display with the tubing mentioned above). Some people use brazing rod for the radials, but they are a little hard to work in my opinion. The ends of the radials are bent into a hook and inserted into the screw holes on the SO-239. Once seated and positioned correctly (see top view inset to *Figure 11-4*) they are soldered in place. Use 60/40 or 50/50 lead/tin radio-electronic solder, NOT plumber's solder.

Mounting details for the ground plane antenna are shown in *Figure 11-5*. Two methods are shown. In *Figure 11-5A* the SO-239 is mounted on an L-bracket that is screwed to a wooden lumber support. If the L-bracket is made of steel, copper or brass, then the SO-239 can be soldered to the bracket at the same time the radials are soldered to the SO-239 (makes it easier).

The other method uses a piece of PVC pipe to hold the antenna (the radials are not shown for simplicity). A sectioned view in *Figure 11-5B* shows the PL-259 on the end of the coaxial cable screwed onto the

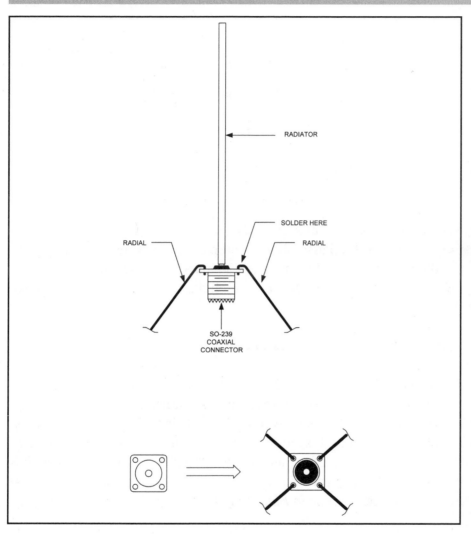

RADIATOR

SOLDER HERE

RADIAL

RADIAL

SO-239
COAXIAL
CONNECTOR

Figure 11-4.
Simple homebrew
groundplane.

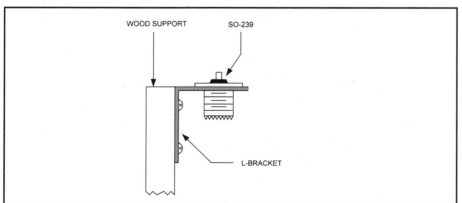

WOOD SUPPORT

SO-239

L-BRACKET

Figure 11-5A.
L-bracket mounting
a VHF/UHF vertical
or Ground Plane.

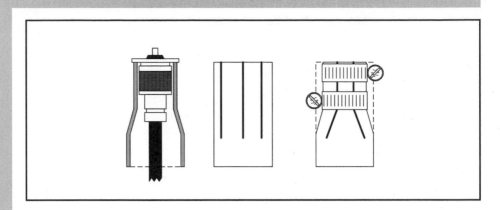

Figure 11-5B.
PVC pipe method
for mounting a
VHF/UHF vertical or
Ground Plane.

SO-239, with the PVC pipe around it. Also shown in *Figure 11-5B* is the preparation and final assembly. To prepare the PVC use a small saw to slit the end as shown. Insert the antenna and feedline, and then place a pair of hose clamps over the end. Cinch the hose clamps down to make the assembly stable.

Vertical Collinear J-Pole Antenna

Figure 11-6 shows a collinear J-pole antenna. Like the ground plane in *Figure 11-5* it has an omnidirectional azimuthal pattern, but provides some amount of signal gain over the ground plane. The antenna consists of two 5/8-wavelength sections ("A") separated by a phasing harness ("B"). At the bottom of the antenna is a matching section that matches the impedance of the coaxial cable to that of the antenna base. The dimensions of the antenna in *Figure 11-6* are:

$$A = \frac{186,919}{F_{MHZ}} \text{ millimeters} \qquad (11\text{-}3)$$

$$B = \frac{32,569}{F_{MHZ}} \text{ millimeters} \qquad (11\text{-}4)$$

$$C = \frac{67,970}{F_{MHZ}} \text{ millimeters} \qquad (11\text{-}5)$$

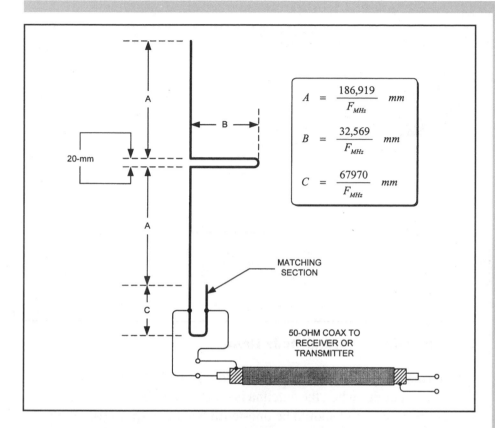

$$A = \frac{186,919}{F_{MHz}} \; mm$$

$$B = \frac{32,569}{F_{MHz}} \; mm$$

$$C = \frac{67970}{F_{MHz}} \; mm$$

20-mm

A

B

A

MATCHING
SECTION

C

50-OHM COAX TO
RECEIVER OR
TRANSMITTER

Figure 11-6.
Collinear antenna.

The antenna can be constructed of No. 10 solid copper wire on the cheap, or small-diameter brass tubing of the sort mentioned above for the ground plane. Attach the coaxial cable about 2.5 cm from the bottom of the matching section and measure the VSWR. If it is satisfactory, then solder it in place. Otherwise, move the connection point while monitoring the VSWR until as close to 1:1 as possible is obtained.

418 MHz Stick Antenna

Figure 11-7 shows an antenna that is popular with 418 MHz telemetry users. It can be made simply from readily available materials. The direction of maximum radiation and reception is along length "A" in the diagram. The dimensions shown are for 418 MHz, but scaling can be done. However, the dimensions and number of turns for the inserted coil must be determined experimentally. The ground plane disk is connected to the coaxial cable shield, and should be about 100 mm diameter.

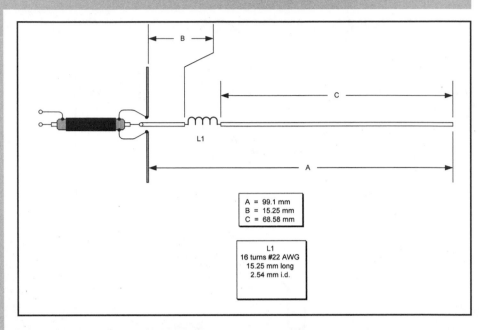

A = 99.1 mm
B = 15.25 mm
C = 68.58 mm

L1
16 turns #22 AWG
15.25 mm long
2.54 mm i.d.

Figure 11-7.
Stick antenna.

Omnidirectional Normal Mode Helix

The normal mode helix antenna shown in *Figure 11-8* produces an omni-directional pattern when the antenna is mounted vertically. The diameter (D) of the helical coil should be one-tenth wavelength (λ/10), while the pitch (i.e., S, the distance between helix loops) is one-twentieth wavelength (λ/20). An example of the normal mode helix is the "rubber ducky" antenna used on VHF/UHF two-way radios and scanners.

Axial Mode Helical Antenna

An axial mode helical antenna is shown in *Figure 11-9*. This antenna fires "off-the-end" in the direction shown by the arrow. The helix is mounted in the center of a ground plane that is at least 0.8λ across. For some UHF frequencies some builders have used aluminum pie pans for this purpose. The helix itself is made from either heavy copper wire (solid, not stranded) or copper or brass tubing. The copper tubing is a bit easier to work. The dimensions are:

$$D \approx \lambda/3$$
$$S \approx \lambda/4$$
$$\text{Length} \approx 1.44\lambda$$

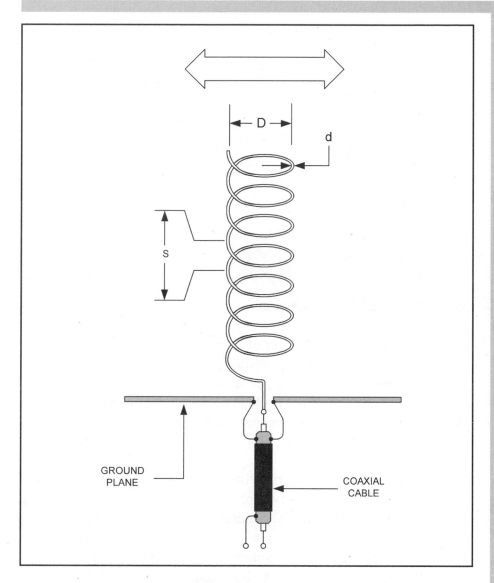

Figure 11-8.
Normal mode
helical antenna.

A "rule of thumb" for the circumference is that maximum gain is obtained when circumference C is:

$$C = 1.066 + [(N - 5) \times 0.003] \qquad (11\text{-}6)$$

The actual gain equation is a bit complicated, but can be calculated in antenna modeling programs or even in simple Excel™ spreadsheet programs.

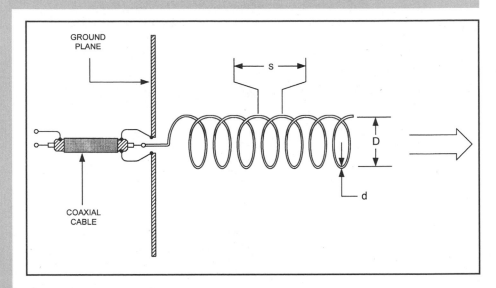

Figure 11-9.
Axial mode helical
antenna.

Coffee Can Microwave Antenna

When I was in college a number of engineering students were making 2.145-GHz microwave antennas using 2-pound coffee cans. Their purpose was to pick up the multipoint distribution service (MDS) microwave signals that carried movie channels to apartment house and hotel customers...until the dean stopped it (it was illegal). Some hams also use this form of antenna. *Figure 11-10* shows the basic construction.

A hole is drilled in the center of the bottom of the can for a coaxial connector suitable to microwave applications. Most of the engineering students used an ordinary BNC connector, although other forms might be more suitable. A small hairpin pickup loop is connected to the center conductor of the connector, and soldered to the bottom of the can beside the connector. The loop should be about quarter wavelength long. Some people used a small microwave half wavelength dipole element at that point.

This antenna assumes that you want to route a signal to a receiver. Mount a low-noise preamplifier on the bottom of the can, before the signal is sent down the coaxial cable to the receiver. The lid can be used to weatherproof the antenna, although a small amount of signal loss will occur.

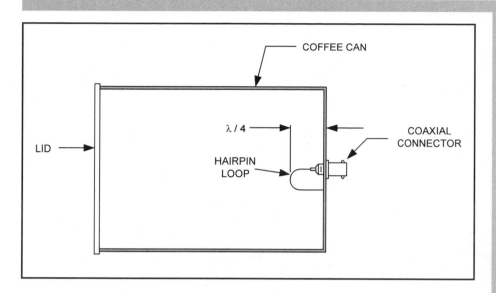

COFFEE CAN

$\lambda / 4$

COAXIAL
CONNECTOR

LID

HAIRPIN
LOOP

Figure 11-10.
Coffee can
microwave horn.

Chapter 12
FINDING BEARINGS ON
EARTH—THE GREAT CIRCLE
CALCULATING RADIO ANTENNA BEARINGS

Chapter 12
Finding Bearings on Earth — The Great Circle
Calculating Radio Antenna Bearings

If you use a directional antenna (especially unidirectional antennas such as the yagi or quad beam) in your RadioScience Observing, then it might be nice to know the direction in which to point the darn thing. The trick is to know the *great circle bearing* between your location and the other station's location. That bearing is calculated from some simple spherical trigonometry using a handheld calculator or a computer program. Before talking about the math, however, we need to establish a frame of reference that makes the system work.

Latitude and Longitude

The need for navigation on the surface of the Earth caused the creation of a grid system to uniquely locate points on the surface of our globe. *Figure 12-1* shows how this system works. *Longitude* lines run from the north pole to the south pole; i.e., from north-to-south. The reference point (longitude zero), called the *prime meridian*, runs through Greenwich, England (*Figure 12-2*). The longitude of the prime meridian is 0 degrees. Longitudes west of the prime meridian are given a plus sign (+), while longitudes east of the prime are given a minus (-) sign. If you continue the prime meridian through the poles to the other side of the Earth it has a longitude of 180 degrees; thus, the longitude values run from -180 degrees to +180 degrees, with ±180 degrees being the same line.

The observatory at Greenwich is also the point against which relative time is measured. Every 15-degree change of longitude is equivalent to a one hour difference with the Greenwich time. To the west, subtract one

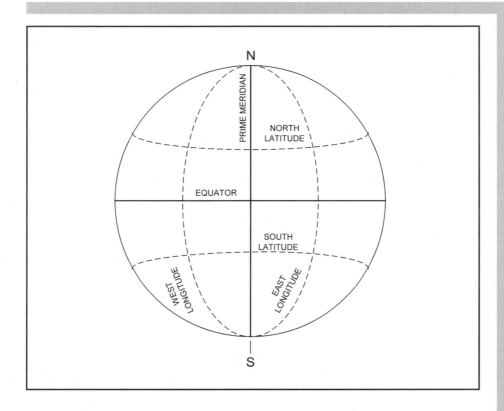

Figure 12-1.
The latitude and
longitude grid for
Earth.

hour for each 15 degrees and to the east add one hour for each 15 degrees. Thus, the time on the east coast of the United States is -5 hours relative to Greenwich time. At one time, we called time along the prime meridian *Greenwich mean time* (GMT), also called *Zulu time* to simplify matters for CW operators.

Latitude lines are measured against the equator (*Figure 12-2*), with distances north of the equator being taken as positive, and distances south of the equator being negative. The equator is 0 degrees latitude, while the north pole is +90 degrees latitude and the south pole is -90 degrees latitude.

Navigators long ago learned that the latitude can be measured by "shooting" the stars and consulting a special atlas to compare the angle of certain stars with tables that translate to latitude numbers. The longitude measurement, however, is a bit different. For centuries sailors could measure latitude, but had to guess longitude (often with tragic results).

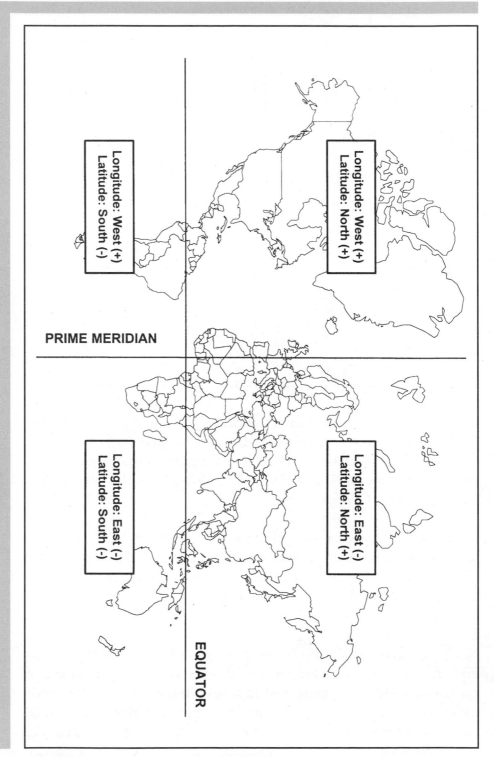

Figure 12-2.
World divided in
lat-long zones.

In the early 18th century, the British government offered a large cash prize to anyone who could design a chronometer that could be taken to sea. By keeping the chronometer set accurately to Greenwich mean time, and comparing GMT against local time (i.e., at a time like high noon when the position of the Sun is easy to judge), the longitude could be calculated. If you are interested in this subject, then most decent libraries have books on celestial navigation.

The Great Circle

The shortest distance between two points is a straight line, right? Nope, not on a globe. On the surface a globe, a curved line called a *great circle path* is the shortest distance between two points. This path can cause some interesting anomalies. For example, I live on a latitude that is close to the latitude of Lisbon, Portugal (in which case, why do *they* get the good weather?). Given that fact, one might assume that I would point my beam due east, at a bearing of 90 degrees from true north. If I did that, I might hear Portuguese voices coming over the receiver, but they would be from the west coast of Africa, close to Angola (a former Portuguese colony).

Figure 12-3 shows the basic problem for calculating antenna bearings. Consider two points on a globe: "A" is your location, while "B" is the other station's location. The distance "D" is the great circle path between "A" and "B."

The great circle path length can be expressed in either degrees or distance (e.g., miles, nautical miles or kilometers). To calculate the distance, it is necessary to find the difference in longitude (L) between your longitude (LA) and the other guy's longitude (LB): L = LA - LB. Keep the signs straight. For example, if your longitude (LA) is 40 degrees, and the other guy's longitude (LB) is -120 degrees, then L = 40 - (-120) = 40 + 120 = 160. The equation for distance (D) is:

$$\cos D = (\sin A \ \text{x} \ \sin B) + (\cos A \ \text{x} \ \cos B \ \text{x} \ \cos L) \qquad (12\text{-}1)$$

Where:

> D is the angular great circle distance
>
> A is your latitude
>
> B is the other station's latitude.

To find the actual angle, take the arccos of Equation 12-1, i.e:

$$D \quad = \quad \arccos{(\cos D)} \qquad\qquad (12\text{-}2)$$

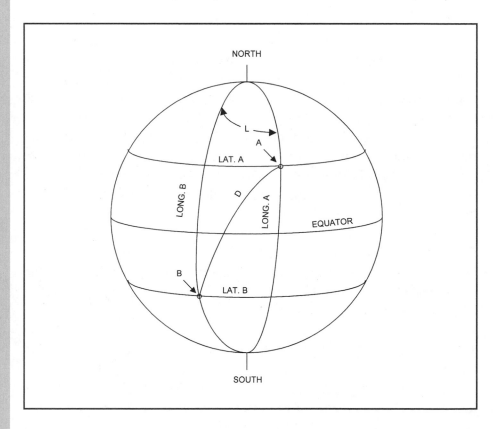

Figure 12-3. Definitions of calculation variables.

In the next equation you will want to use D in angular measure, but later on will want to convert D to miles. To do that neat trick, multiply D in degrees by 69.4. Or, if you prefer metric measures, then D x 111.2 yields kilometers. This is the approximate distance in statute miles between "A" and "B."

To find the bearing from true north, then work the equation below:

$$C = \arccos \left[\frac{\sin B - (\sin A \times \cos D)}{(\cos A \times \sin D)} \right]$$

(12-3)

Now, for the rub: *This equation won't always give you the right answer unless you make some corrections.*

The first problem is the *same longitude error*; that is, when both stations are on the same longitude line. In this case, L = LA - LB = 0. If LAT A > LAT B, then C = 180 degrees, but if LAT A < LAT B, then C = 0 degrees. If LAT A = LAT B, then what's the point of all these calculations?

The next problem is found when the condition $-180° \leq L \leq +180°$ is not met; when the absolute value of L is greater than 180° (ABS(L) > 180°). In this case, either add or subtract 360 in order to make the value between ±180°:

If L > +180, then L = L - 360

If L < -180, then L = L + 360

One problem seen while calculating these values on a computer is the fact that in BASIC the sin(X) and cos(X) cover different ranges. The sin(X) function returns values from 0° to 360°, while the cos(X) function returns values only over 0° to 180°. If L is positive, then the result of Equation 12-3, bearing C, is accurate, but if L is negative then the actual value of C = 360 - C. I ran across this problem when trying to compare the results of calculations from New York, NY (40.43N, 77W) to Japan and points in Australia. I had expected some bearings in the northwesterly direction because of the great circle map published in older editions of the *ARRL Antenna Book*. Oops! After doing a bit more research, I found the error and added the test below to my program:

IF L < 0 then

L = 360 - L

Else L = L

End if

Another problem is seen whenever either station is in a high latitude near either pole (±90°), or where both locations are very close together, or where the two locations are antipodal (on opposite points on the Earth's surface). According to Hall (1973), the best way to handle these problems is to use a different version of Equation 12-3 that multiplies by the cosecant of D (csc(D)), rather than dividing by sine of D (sin(D)).

There is a program on the CD-ROM that calculates the Great Circle Bearings.

Acknowledgment

My thanks to the ARRL Technical Department for aid in locating Hall's article, as well as other material on the problem of bearing calculations.

Reference

Jerry Hall, K1PLP (1973). "Bearing and Distance Calculations by Sleight of Hand," *QST*, August 1973, pp. 24-26.

Chapter 13
VIBRATION DETECTORS AND SEISMOGRAPHS

Chapter 13
Vibration Detectors and Seismographs

A number of scientific and engineering investigations measure vibration signals. For example, engineers can often characterize a metal plate or beam by placing a vibration sensor on it, and then giving it an impact. I've seen one case where they placed strain gauge sensors on a metal beam (*Figure 13-1*) and then whacked the other end with a "dead blow" hammer (a hard rubber hollow mallet filled with BB shot). The idea behind using the "dead blow" hammer is to prevent bouncing of the hammer from creating more than one blow. In another case, sensors were placed at critical points on a bridge, and then a small explosive charge was ignited at the other end. The charge was on the order of a cherry bomb, so caused no damage. In both cases, the idea was to cause an impact, and then record the vibrations in the structure that result from it.

If you saw the movie *Jurassic Park*, then you saw another use of vibration sensors. A shotgun shell blank was held in a rig against the ground and exploded. The seismic waves were recorded on a laptop computer and processed to show the all-too-realistic image of a raptor skeleton.

Oil exploration is done the same way. An explosion is set off at a site, and a cluster of vibration sensors around a perimeter are used to sense the vibrations. From their data the geologists can construct an image of the underlying structure, and from it predict where oil might be found.

Figure 13-1. Using sensor to measure resonant vibration profile on beam.

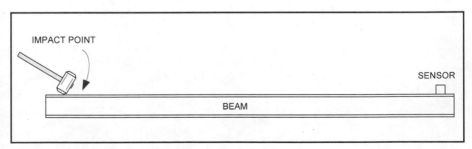

The idea behind any sensor is to find some transducible property that responds to the event being measured. A number of things can be used. For example, the strain gauge consists of a thin resistive wire stretched across a membrane or diaphragm. When the wire is deformed, its dimensions change, so its resistance will also change. This phenomenon is called *piezoresistivity*. When the strain gauge resistor is used in a Wheatstone bridge, then a sensitive measure of the deformation caused by vibration can be obtained.

Another transducible property for vibrations is the inductor shown in *Figure 13-2A*. The inductance of the coil of wire is determined by the number of turns, diameter of the coil, the length of the coil and the nature of the core. If a ferrite or powdered-iron core is used, then a large increase of inductance occurs when the core is slipped into the coil form. "Slug-tuned" inductors are used in radio circuits. A threaded core is adjusted to be more or less inside the coil, depending on the value of inductance required. Vibration sensors can be made by placing the core on a spring, pendulum or some other means of translating the motion caused by vibration into motion of the core...and therefore a change of inductance. In *Figure 13-2B* a pendulum is used to move the core in and out of the coil form.

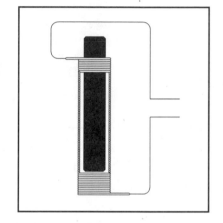

Figure 13-2A.
Coil and core sensor.

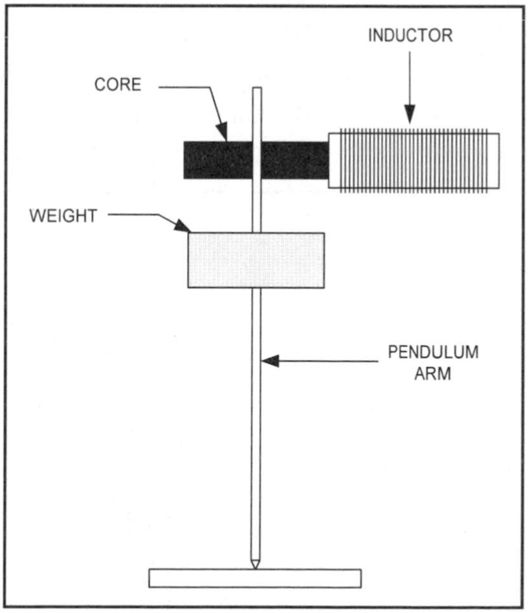

Figure 13-2B.
Actuation of core.

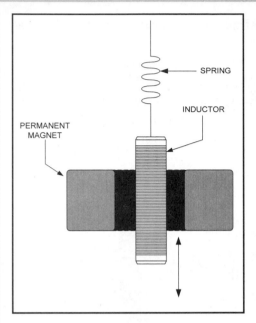

There is a class of motion sensors that depend on the fact that a magnetic field and coil in motion relative to each other cause a current to be induced into the coil. It doesn't matter whether the coil moves or the magnet moves, so long as there is relative motion.

Figure 13-3 shows a crude form of *permanent magnet moving coil* (PMMC) vibration sensor. A horseshoe magnet is positioned such that an inductor can move inside of its field. The coil, which may have many, many turns, is connected to a spring. When vibration is sensed, the coil moves up and down against the spring, causing a current to be induced because of the magnetic field.

Figure 13-3.
Large vibration sensor using permanent magnet and moving coil (PMMC).

A more familiar example is something that you might not at first see as a vibration sensor. The ordinary radio or television loudspeaker (*Figure 13-4*) will do this neat trick. The PMMC loudspeaker consists of a fixed permanent magnet (usually made of Alnico) and a lightweight moving coil that is attached to the paper speaker cone. When an electrical cur-

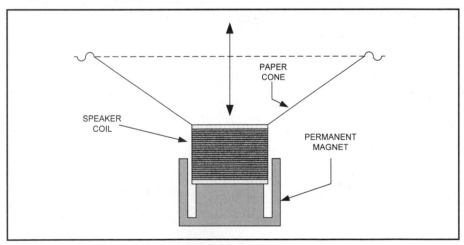

Figure 13-4.
Loudspeaker vibration sensor.

rent from the radio output stage flows in the coil it will produce a magnetic field that either attracts or repels the permanent magnet's field, depending on the polarity of the audio signal. Thus, as the polarity of the audio signal switches back and forth, the coil and cone move in and out.

The same loudspeaker that can be used as a radio output device can also be used as a microphone. Intercom sets that allow talk-back use this phenomenon. When sound vibrations impinge the cone, it will move the coil relative to the magnet, causing a current to be induced into the coil. This tiny current can be amplified and used as an audio signal.

Note that word in the previous paragraph: *vibration*. If the vibrations being measured can be coupled to the speaker cone, then the speaker will act as a vibration sensor. I've seen speakers placed flat against metal plates for the purpose of recording vibrations. One science fair student cemented a plastic drinking straw against the bottom of the speaker cone, and used it to couple to the vibration source. Essentially, the kid made a large-scale "spike microphone."

Linear Differential Voltage Transformers (LDVT)

Figure 13-5 shows a special form of vibration or displacement sensor called a *linear differential voltage transformer* (LDVT). It consists of three inductors (L1 through L3). Inductor L1 is excited by an AC signal, which is magnetically coupled to secondary coils L2 and L3. When the core is exactly midway between L2 and L3, the currents flowing in them will be identical. The dots indicate the phasing of the two secondary coils. Because of the connection of L2 and L3, the currents are series opposing; thus, when the core is midway, the currents are equal and opposite, so null to zero. But when the core is offset in one direction or another, then one of the coil currents will predominate.

The amplitude of the output signal (expressed as a voltage) gives us an indication of the magnitude of the core shift, while the polarity tells us the direction. When the core is coupled to something like a pendulum or

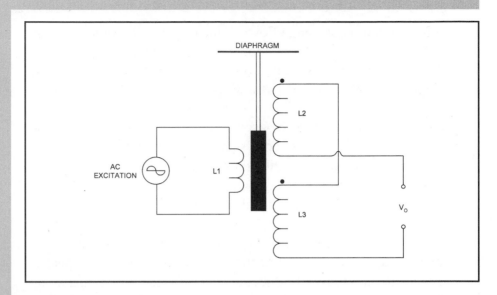

Figure 13-5.
LDVT sensor.

diaphragm, then the vibrations received will move the core in an out of the coil-pair, causing an AC output signal to appear.

Differential Capacitor Sensors

Capacitance exists whenever two metallic objects are in close proximity to each other, and are separated by an insulating material (dielectric). Such a device is called a *capacitor*. Capacitance is a measure of the capacitor's ability to store an electrical charge in an electrical field set up between the plates. The value of capacitance is proportional to the area of the metal surfaces and a property of the dielectric called the *dielectric constant* ($\varepsilon = 1$ for dry air), and inversely proportional to the distance between the metal surfaces (in other words, the closer together they are, the higher the capacitance).

Capacitors are used in a variety of electronic applications. But in this particular case we are interested in a class of capacitors that are variable, and can be made to vary under the influence of vibration.

Differential Capacitors. This class of variable capacitor actually consists of two variable capacitors actuated by the same shaft or other mechanical device. They are configured such that one capacitance increases its capacitance while the other decreases its capacitance in the same manner. If you were to measure the total capacitance across the two capacitors, then you would find that the net capacitance does not change even though the values of the two capacitor sections do.

Figure 13-6 shows a vibration sensor that works on the differential capacitance phenomenon. The two capacitors are formed by two metal cylinders (C1 and C2), separated by a small dielectric gap of air or other material. These cylinders share a metal plunger that is inside and axially concentric with them. An insulating sleeve prevents the plunger (the "common plate" of the differential capacitor) from shorting out against either cylinder. When the plunger is equally inside both cylinders, favoring neither, then their respective capacitances are equal. But if the actuating arm moves, then the plunger will move more deeply into one cylinder and out of the other. As a result, the ratio of the two capacitances changes.

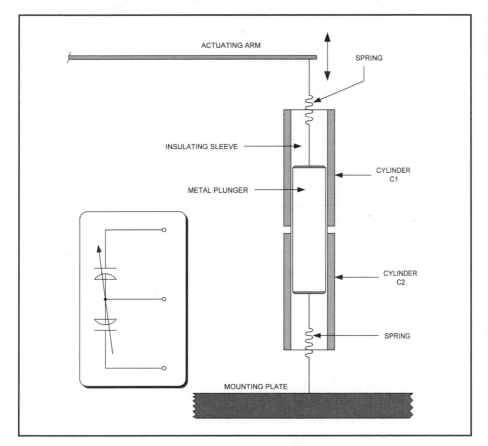

Figure 13-6. Cylindrical differential capacitor sensor.

The inset in *Figure 13-6* shows the equivalent circuit schematic for the differential capacitor vibration/displacement sensor.

A circuit for using a differential capacitance sensor is shown in *Figure 13-7*. The differential capacitor (C1A and C1B) is connected as two arms of a Wheatstone bridge. The remaining two arms are resistors R1 and R2. It is the nature of this type of bridge circuit that output voltage V_O will be zero when the ratios of the capacitive reactances and resistances are equal:

$$\frac{X_{C1A}}{X_{C1B}} = \frac{R1}{R2} \qquad \text{for } V_o = 0 \qquad\qquad (13\text{-}1)$$

If R1 = R2, the output voltage will be zero when the differential capacitor is balanced; i.e., C1A = C1B. When vibration or other motion displaces the plunger, however, C1A ≠ C1B so the bridge is unbalanced and V_O is non-zero. The amplitude of V_O depends on the amplitude of the mechanical displacement of the plunger.

Other forms of circuit can be used with a differential capacitive sensor. For example, the two capacitors (C1A and C1B) can be used to control the frequency of radio frequency oscillators. When C1A = C1B, then the two frequencies would be equal, but a change in that equality will force the frequencies apart. One frequency will rise, while the other will fall. If

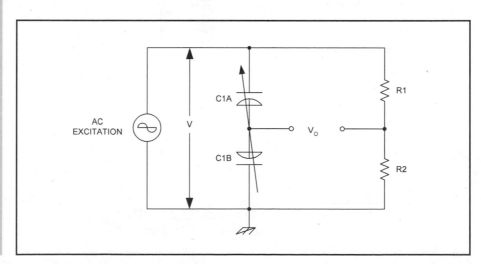

Figure 13-7.
Differential capacitor
in bridge circuit.

the two frequencies are combined in a nonlinear mixer, then the resultant heterodyne beat note will be proportional to the deflection of the plunger. A frequency-to-voltage (F/V) converter or other form of circuit can be used to produce the analog waveform.

Another approach is shown in *Figure 13-8*. It is derived from a sensor called the Shackleford-Gunderson seismometer. In this type of sensor circuit the common plate of the capacitor is excited by a 4-to-6 MHz RF signal, and the other two are used as "receive antennas." When the detected and integrated outputs of these receive antennas are combined, the resultant signal is proportional to the vibration.

Still another approach is to charge the capacitors through a high-value resistor with a DC source, and then use an electrometer to measure the voltage across the capacitors. The voltage will be proportional to the charge, which in turn is set by the capacitance. The result is that the two DC voltages will vary according to the position of the plunger. Note: *Electrometers* are amplifiers with extremely high input impedances, so can be used to measure the charge developed across small-value capacitors.

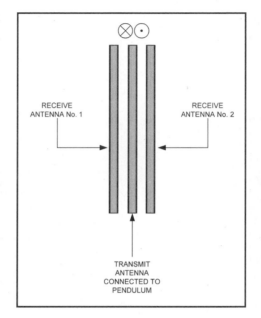

Figure 13-8. Differential receive antenna capacitive sensor.

Another form of differential capacitor vibration sensor is shown in *Figure 13-9*. This device uses a pendulum to move a metal disk between two sections of the capacitor. One inset shows the schematic symbols for C1A/C1B, while the other shows a top view of the capacitor/pendulum. The stator plates of C1A and C1B are made with bits of sheet metal, or blank unetched printed-circuit board stock (which might be easier to work).

Two sets of plates are used for each capacitor, and they are connected together. As the pendulum bob swings back and forth it will "shade" more or less of each set of stator plates, depending on the amplitude and direction of the swing. Similar circuit strategies as above will also work for this sensor.

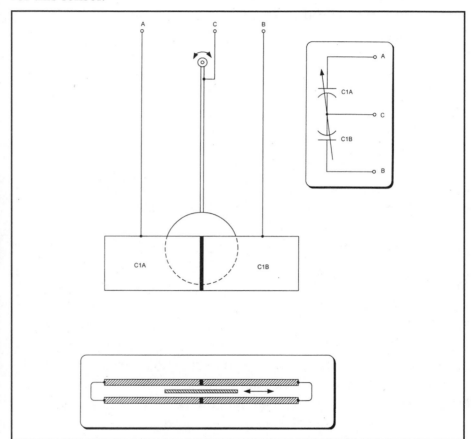

Figure 13-9.
Pendulum differential capacitor sensor.

Lehman-type Seismometer Project

The Lehman seismometer was designed by Dr. James D. Lehman of the Department of Physics at James Madison University in Harrisonburg, Virginia. It was the subject of the "Amateur Scientist" column written by Jearl Walker in *Scientific American* (July 1979, p. 152ff). This seismometer is capable of detecting earthquakes of 4.8 or more on the Richter scale in the United States, and 6 or more in other areas of the world.

The plan (top) view is shown in *Figure 13-10*, while a side view is shown in *Figure 13-11*. A long (75 cm or more) pendulum boom made of 5/16" (8 mm) stock is suspended from an upright section (*Figure 13-12*) such that its "knife edge" butt end rests against a strike plate on the upright section. The last 20 cm or so of the boom is threaded to accept a 5/16-20 or finer hex nut. When side-to-side vibrations, characteristic of an earthquake, are present, the pendulum arm will swing back and forth in the horizontal plane. A pair of stops (*Figure 13-13*) are used to prevent overtravel of the pendulum boom arm. The components of the seismometer are mounted on a flat base plate. This plate can be made of aluminum, wood or plastic, but it must be level and stable. Do not use thin material (less than 1.5 cm) for the base.

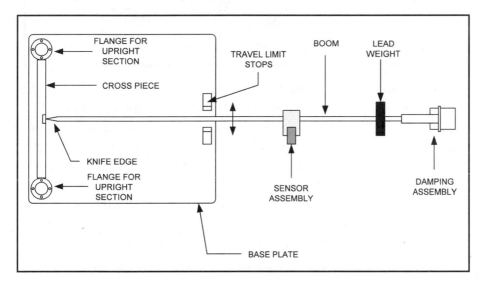

Figure 13-10.
Lehman seismometer
(top view).

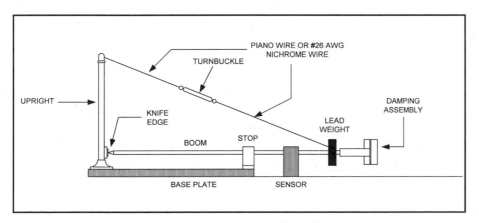

Figure 13-11.
Lehman seismometer
(side view).

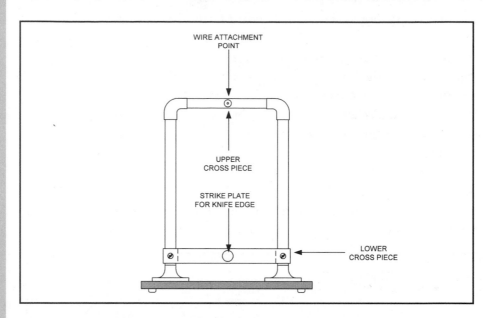

Figure 13-12.
Upright view.

The upright segment is made from copper plumbing pipe, right-angle joints, and a pair of universal flanges for mounting to the base plate. The overall height is about 46 cm. A lower crosspiece holds the strike plate, while the upper crosspiece holds a special wire attachment point fixture (about which, more shortly).

In *Figure 13-11* you will see the side view. The wire suspending the boom is either piano wire or #26 AWG nichrome wire. A small turnbuckle is used to make the wire taught and level the boom.

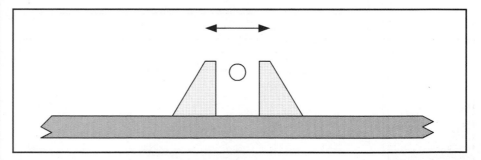

Figure 13-13.
Mechanical limit
stops.

The upper end of the suspension wire is tied off to the upper crosspiece (*Figure 13-14*). The Walker article used an oil burner furnace nozzle to precisely position the suspension wire.

Detail in plan view of the lower crosspiece is shown in *Figure 13-15*. The crosspiece can be machined from a bar of aluminum or brass, although some have made it out of plastic (not recommended, I am told). The ends of the crosspiece are beveled to allow it to be mounted to the ver-

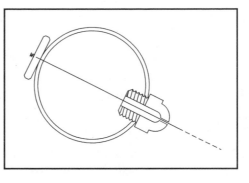

Figure 13-14.
Taut wire stanchion.

tical upright members. In the center of the crosspiece is the strike plate assembly. Detail for the strike plate itself is shown inset. A 1/4-inch carriage bolt or other flathead machine bolt is used to make the strike plate. It is cut off from the threaded segment, leaving only the head and the unthreaded portion of the shank. The head is machined flat and polished. This makes a nice, low-friction surface for the knife-edge end of the pendulum boom.

A hole is drilled through the crosspiece just slightly larger than the strike plate. A 1/4-20 (or finer) hex nut is soldered or glued to the backside of the crosspiece such that its hole is centered over the hole drilled in the crosspiece. A matching hex-head bolt is threaded into the nut, so that it sits against the back of the strike plate. When this bolt is adjusted, it will adjust the position of the strike plate.

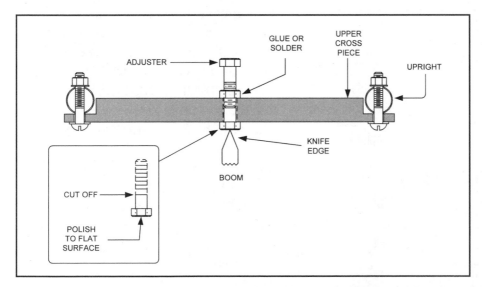

Figure 13-15.
Strike plate and
knife edge detail.

A sensor assembly is mounted on the boom arm (refer again to *Figures 13-10* and *13-11*). It is used to detect motion of the boom and translate it into an electrical signal that can be recorded on paper or stored in a computer. Also on the boom arm is a lead weight to make the boom act like a pendulum, and a damping assembly. The damper is needed to damp out oscillations of the pendulum in order to spread out the frequency response. Without the damping assembly the vibrations closest to the pendulum's natural period will cause the greatest response. The frequency response is thus peaked at a specific frequency, rather than being broader.

Detail for the sensor assembly is shown in *Figure 13-16*. A plan view is shown in *Figure 13-16A*, and a side view is shown in *Figure 13-16B*. The sensor's "transducible event" results from the fact that a magnetic field moving relative to a coil of wire will induce a current in that wire. In *Figure 13-16A* you will see a strong horseshoe magnet mounted on the boom arm of the pendulum. The magnet is fixed with a wing nut connector of the sort used to fasten chemistry and optical apparatus.

Inside the curve of the horseshoe magnet is a coil of wire. The coil should be several centimeters in diameter, and consist of about 8,000 to 12,000 turns of #36 AWG (or so) enameled wire. The former for the coil can be constructed of wood or plastic. A slow-speed drill is used to scramble-wind the wire onto the form.

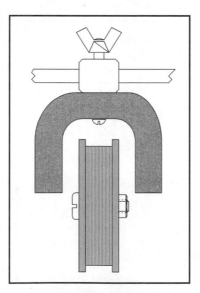

Figure 13-16A.
Sensor detail.
(plan view)

Winding the coil can be a dangerous operation, and you take your own responsibility if you attempt it. Make sure the drill has a very slow speed (most variable speed drills will do this). Fix the drill and the form in a vise, or other fixture. The wire should be on a dowel, mounted offset from the drill and form, but positioned so that it can be unwound without kinking. WEAR PROTECTIVE SAFETY GOGGLES WHEN WINDING THE COIL. Injury to your eyes could occur if the wire breaks, or if you do something wrong.

The mounting of the coil and magnet is shown (side view) in *Figure 13-16B*. The mounting assembly can be made from either wood or plastic, or any convenient material.

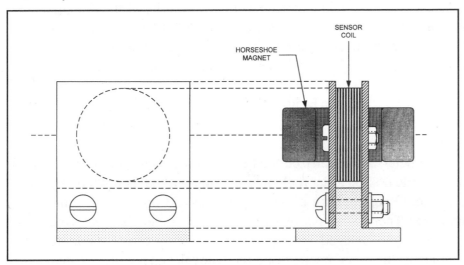

Figure 13-16B.
Sensor detail.
(side view)

The damping assembly is shown in both plan and side views in *Figure 13-17*. The lead weight is made of plumber's lead, and is fixed to the threaded end of the boom with a pair of hex nuts. The period of the pendulum can be adjusted by the position of the weight, if needed.

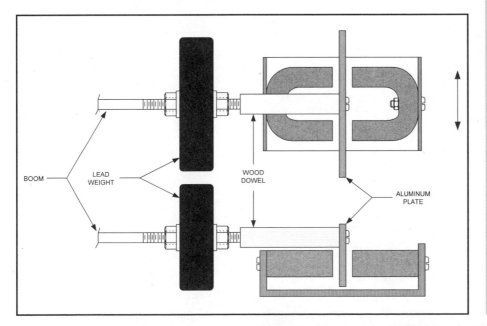

Figure 13-17.
Damper detail.

The damping action is created by swinging a metal plate through the magnetic fields of opposing horseshoe magnets. When the pendulum moves, eddy currents are set up in the aluminum plate, and these in turn create a magnetic field that is opposite the polarity of the field that created the current. This is a magnetic version of "rust on the door hinges" to keep it from swinging for a long time after excited by vibration. The aluminum plate is fixed to the pendulum by a wooden or plastic dowel. One end of the dowel is threaded to fit the thread on the 8-mm boom end. A wood screw can be used to fasten the aluminum plate to the other end of the dowel.

Seismometers are usually buried below grade. Some people place them in a simple box on a flat rock outcropping, but that is not terribly good practice. Others build a concrete pedestal and box to house the seismometer. Thermal insulation surrounding the inner chamber, where the seismometer is located, is used to keep the temperature stable. It is essential that the seismometer be perfectly level, and not tilted. It may take several days to get the seismometer adjusted properly.

Chapter 14
SEARCHING FOR ET FROM
YOUR OWN BACKYARD

Chapter 14

Searching for ET From Your Own Backyard

H. Paul Shuch, Ph.D.
(An Invited Chapter)

Dr. Shuch is Executive Director of The SETI League, Inc. He can be contacted at
P.O. Box 555 Little Ferry, NJ, 07643, or via E-mail at *n6tx@setileague.org*

Perhaps ET is not calling home at all. Maybe he's calling us. As reported in *Satellite Times* (Sept/Oct 1997, page 8), the Search for Extra-Terrestrial Intelligence (SETI), once a multi-million dollar NASA venture, is now in private hands. In 41 countries on six continents, hundreds of dedicated experimenters are now using their backyard SETI stations to seek out that elusive needle in the interstellar haystack. You can't buy a radio telescope at your local Radio Shack store. But if you're reasonably handy with tin-snips, know which end of the soldering iron is the handle, and have a few hundred (to a few thousand) dollars to invest, you too can join the search for our cosmic companions.

What We're Looking For

Our high-technology Earth is surrounded by a telltale sphere of artificial radiation, now extending out to about 50 light years or more, and still traveling outward at the fastest possible speed: the speed of light (about 300,000,000 m/s). This radio, TV, radar and microwave pollution is readily detectable to any local civilization which has radio astronomy. We figure that some of the countless beings living in the light of distant suns may also pollute their radio environment, and we stand a reasonable chance of detecting them.

But don't expect to tune in alien "I Love Lucy." Interstellar signals will be so weak that our eyes and ears will never recognize them. The most we can hope for is order in the cosmic chaos, patterns which could not have been produced by any natural mechanism which we know and understand. These hallmarks of artificiality are evident to computers, and it is your home computer which will sift through the cosmic static in search of ET.

Now, where on the dial should we look? It's highly unlikely that ET honors the FCC's band plans, so we can only guess as to their likely channel lineup. There may well be many good frequencies for SETI, but what they have in common is their ability to pass unimpeded through the interstellar medium. Since the space between the stars is most transparent in the microwave spectrum, that's where we'll start our search.

Satellite TV is broadcast in the microwave region. So are radar, cellular telephone, and much of Earth's telecommunications relay signals. There are also navigation signals from the swarm of Global Positioning Satellites surrounding our planet. If we're going to seek out weak signals from the stars, we need to search in the gaps between our own transmissions. One such interesting gap (there are others) is the resonant frequency of hydrogen atoms, 1420 MHz, and many amateur and professional SETI stations start out there.

What You'll Need: The Antenna

Although other configurations are sometimes used, the hands-down favorite for snagging alien photons is the parabolic reflector, or dish antenna. A 3-to-5 meter diameter dish (that is, 10 to 16 feet) is just about the right size to stand a reasonable chance of SETI success. The classic C-band backyard satellite TV dish is ideal. These have high gain, narrow beamwidth, work over a wide range of frequencies, and are readily available for next to nothing, both new and used. Around the country millions of TV viewers are upgrading to Ku-band Direct Broadcast Satellite (DBS) reception. Its half-meter dishes are very appealing. That leaves millions of C-band BUDs (Big Ugly Dishes) sitting around gathering rust. Many SETI enthusiasts have found neighbors anxious to have these eyesores taken off their hands. And if you're a satellite TV fan, then chances are you already have one. You can also build a dish of your own if you are reasonably handy.

You can use your satellite TV dish to focus 1420 MHz energy, but not its C-band feedhorn. Plan on building or buying a larger tin can to capture these longer wavelengths. A commercial feed which will directly replace your TVRO horn can be purchased for around $150. If you want to use your BUD to watch TV and do SETI in the background, you can mount your SETI feedhorn next to your TVRO one, and multi-task.

What You'll Need: The Low-Noise Preamp

The purpose of a preamplifier is to take an impossibly weak signal from space, and turn it into merely a ridiculously weak one. You used one of these for satellite TV (it may have been called an LNA, LNB, or LNC), but it probably doesn't work on ET's channel. Fortunately, radio astronomy preamps for the desired frequency range are readily available from a number of sources. Price varies from about $50 for a kit preamp up to perhaps $200 for the top-of-the-line, assembled and tested one. The preamp mounts directly to the feedhorn with a coax connector, and drives the coaxial feedline which runs inside to your receiver. You'll also need to run juice from a 12-volt power supply up to your preamp, either through the feedline or on a separate length of lamp cord or speaker wire.

What You'll Need: The Receiver

Once you've amplified your weak alien signal, you need to break it down to audio components which your computer can analyze. This is the job of a microwave receiver. The earliest amateur SETI stations employed ham radio's old standard, the venerable Icom model R-7000 microwave scanner, and its successors, the models IC R-7100 and R-8500. These highly capable receivers are still a good bet if you can find them, though their $2000-plus price tag exceeds the cost of all other parts of your SETI station combined. Fortunately, some less costly alternatives are just beginning to emerge.

For years ham radio operators have been converting microwave signals down to frequency regions which their existing shortwave receivers can process, and SETI is no exception. For just over a hundred dollars in kit form (twice that if already assembled), you can today buy a downconverter which will shift the most interesting radio astronomy frequency down to the popular 2-meter band, for reception in your existing VHF rig (see *Figure 14-1*). *Figure 14-2* shows the use of a receiver such as the Icom 7100 in an amateur SETI system.

And by adding a $100 2-meter SSB receiver kit to that downconverter, enterprising experimenters have been building their own complete SETI receivers for a small fraction of the cost of commercial units. We hope such packaged special-purpose receivers will come on the market as manufacturers recognize the market potential of SETI.

Lately, receivers-on-a-card are all the rage. For example, Rosetta Labs of Australia makes its *WinRadio* scanning receiver card to plug directly into the motherboard of your personal computer. Though the stock WinRadio is not ideal for the SETI application, look for Rosetta Labs and other vendors to come out with special-purpose SETI receiver cards in the months ahead. Integrated directly into the computer (see below), they promise ever improving performance at significantly reduced cost.

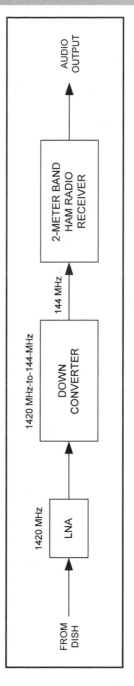

Figure 14-1.
SETI receiver set-up for 1420-MHz using a ham radio 2-meter receiver.

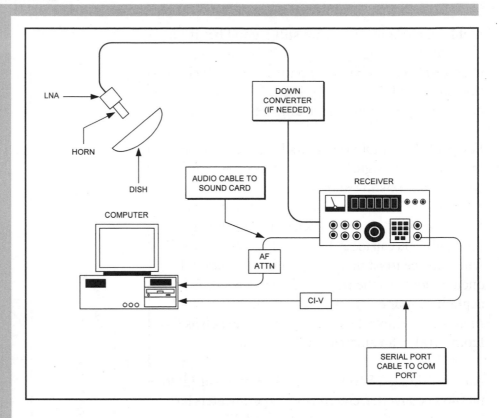

Figure 14-2.
Amateur SETI
station

What You'll Need: The Computer

The purpose of the SETI computer is to run the software which recognizes ET amid the cosmic din. A good bit of number-crunching power is required. The technique is called *Digital Signal Processing*, or DSP, and is the one part of the SETI task which has grown in power at an amazing rate. Raw computer horsepower seems to double every year or so, which means today's home computers are 1000 times more powerful than those of just ten years ago, and 1,000,000 times more powerful than those of two decades past!

We start by breaking down the receiver's audio into digital "ones and zeroes," using a circuit called an Analog-to-Digital Converter, or ADC. There's a very capable ADC in your garden-variety $29 sound card, and that's what most of us are using.

DSP software comes in a variety of flavors, with the most popular varieties being shareware for the DOS and Windows environments. As for the computer on which this software runs, a high-speed Pentium™ is nice, but not essential. Many a SETI enthusiast has used the old 486, which his or her Pentium recently replaced, as a dedicated signal processing machine. And a few SETIzens have even resurrected their old 386 and 286 machines for DSP use. The rule seems to be, any computer you can get your hands on will be more sensitive than your own eyes and ears in separating the alien wheat from the cosmic chaff.

Putting It All Together

All the bits and pieces can be a tad intimidating, but you won't be going it alone. The SETI League is the world's leading grass-roots SETI organization, with hundreds of members in dozens of countries on six continents, and growing. Our website (http://www.setileague.org/), technical manuals and volunteer regional coordinators have already helped hundreds of individual experimenters to get their stations up and running, and stand ready to assist you as well. SETI League members come from all professions, educational levels and walks of life. We share a common curiosity about the beyond, as well as a conviction that we can make a difference.

What We've Heard So Far

Organized SETI has been going on for nearly forty years. About once or twice a year, we detect something strange, a signal which we just can't explain away. Unfortunately, none of these tantalizing candidate signals has yet proven conclusive. SETI demands the most stringent level of proof, if it is to answer the fundamental question which has haunted humankind since first we realized that the points of light in the night sky are other suns: Are We Alone?

The granddaddy of all SETI candidate signals was detected at the Ohio State University radio telescope in 1977. It is universally known as the

"Wow!" signal, after the word scribbled in the margin of the computer printout when investigator Dr. Jerry Ehman first noticed it (see *Figure 14-3*). The "Wow!" was even mentioned in an episode of Fox TV's "The X-Files." After over 100 follow-on studies, the "Wow!" has never repeated. But today's amateur SETI stations are *just as powerful* as the Ohio State facility was 21 years ago, when the "Wow!" was detected. Thus it is our hope that, when enough private SETIzens are up and running, the next "Wow!" will prove less elusive.

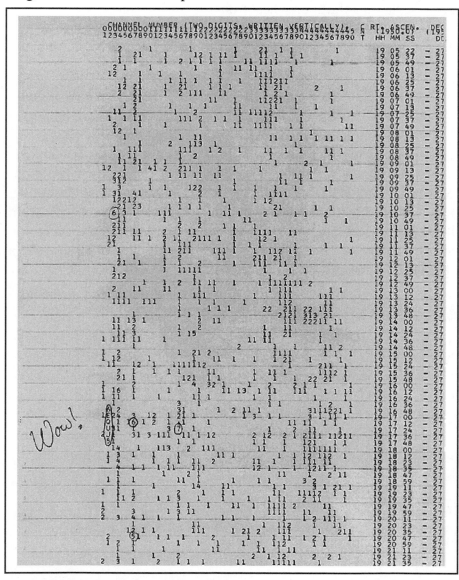

Figure 14-3.
"WOW!" finding.

We've already had a few close encounters. The SETI League's *Project Argus* search of the heavens went on the air in April 1996, initially with just five observing stations (our overall plan calls for 5,000). Only three weeks later, two radio amateurs in England detected

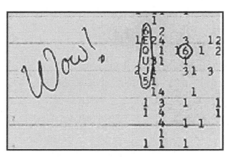

another anomaly. At first glance, this seemed to be just the sort of signal we'd expect from Beyond. It turned out to be a classified military satellite—beyond Earth to be sure, but hardly the ET we were seeking.

Our next interesting signal came from the 1.3-watt beacon transmitter aboard the Mars Global Surveyor satellite, clearly detectable at several million kilometers from Earth. Such detections give us ample encouragement that our systems are up to the task of alien detection. Now all we need is enough participants around the world, coordinated through the Internet, so that no direction in the sky shall evade our gaze. You can be a part of a global net we're stretching to snag that slippery fish in the cosmic pond.

Finding Out More

Check out The SETI League, Inc., a membership-supported, nonprofit educational and scientific organization, on the Internet at http://www.setileague.org/. Leaders in the privatized search for life in space, The SETI League offers technical support, coordination, books, conferences, and a host of related activities for the aspiring SETIzen. Our extensive web site (over 1100 documents totaling more than 36 MBytes, and growing every week) is aimed at the dedicated amateur radio astronomer who's willing to learn. There you'll find sources for the hardware and software discussed above, along with hundreds of pictures showing how others have put their stations together. We have a Technical Manual to help you build, and even our own songbook for those who wish to sing SETI's praises. For membership information, email your postal address to join@setileague.org, or drop us a line at PO Box 555, Little Ferry NJ 07643 USA. We Know We're Not Alone!

Chapter 15
ELECTROMAGNETIC
INTERFERENCE

Chapter 15
Electromagnetic Interference (EMI)

The world is a really hostile place for the types of electronic devices used in RadioScience Observing. Whether you use radio receiving or other devices, consumer electronics equipment, or special purpose electronic instrumentation, it is likely that you will experience *electromagnetic interference* (EMI) at some point. The stuff is all over the place!

Figure 15-1 shows some of the many, many sources of EMI in the normal residential or business building (industrial plants will have even more). A number of these sources have one thing in common: electrical arcing

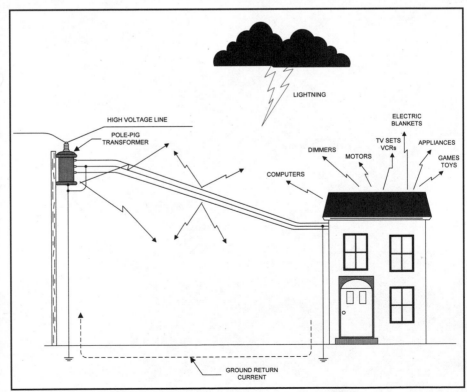

Figure 15-1.
Noise sources
abound...natural
and man-made.

or sparking. Lightning, even when it is far enough away to be out of sight, can produce huge amounts of EMI. The lightning bolt may be either cloud-to-cloud or cloud-to-ground, and consists of about 250,000 amperes of current oscillating back and forth. Any arc produces a large number of harmonics. Indeed, selecting these harmonics was how the old spark gap transmitters (circa *Titanic* era) worked. Lightning will produce significant harmonic energy well into the high-frequency (HF) region of the radio spectrum.

Lightning also produces high-voltage transients on the AC power lines. It does not require a direct hit for lightning to produce these transients. Any electrical current in motion relative to a conductor will induce an electromotive force (emf) into the conductor. When a lightning bolt flashes anywhere nearby, or overhead, an emf will be produced. When electronics were all analog these transients did not usually disrupt electronic circuits unless they were strong enough to create damage (many a silicon power supply rectifier has succumbed to lightning transients on the line). Digital circuits, on the other hand, can be severely disrupted by lightning transients.

The AC power lines are also a source of problem. The standard power system used in North America distributes power via high-voltage lines operated against ground (see *Figure 15-1*). There may be several step-down transformers between your building and the power company generating station. There will be at least one transformer close by your building. It will either be buried or on a pole, as in *Figure 15-1*. The secondary of the transformer is center tapped, with the center tap grounded. The two ends of the transformer form a 240-volt service, while each end and ground form 120-volt services. All three lines are brought into the building.

Several forms of interference are associated with the AC power lines. First, of course, is the fact that there are significant 60 Hz electrical and magnetic fields surrounding the wires from the pole and the wiring inside the building. You can hear this noise when you touch a finger to the audio input of an amplifier hooked to a loudspeaker.

Another source of noise from the AC power lines is arcing. Loose connections are a source of noise problems, and like other forms of arcing will produce harmonic energy a long way above 60 Hz. The power company will repair loose connections on power lines, but you must use an electrician to repair connections inside the building.

A number of appliances in the house will also produce noise. Anything with an electrical motor (air conditioner, refrigerator, furnace, heat pump, etc.) can produce arcing. The problem is at the brushes inside the motor. Like all other forms of arcing, the motor will interfere with circuits and devices operating well into the HF spectrum.

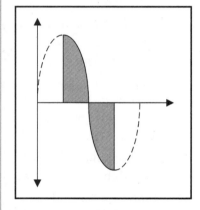

Figure 15-2. Dimmer switches truncate the AC sine wave, causing it to become rich in harmonics.

One class of device that can cause real havoc are those devices that operate using a silicon-controlled rectifier or triac device. Examples include light dimmers, train speed controls, electric blanket controls and so forth. These devices operate to control power by using only a portion of the AC sinewave cycle. In *Figure 15-2* the shaded portion is the power actually applied to the load. The difference between high and lower power settings of the control is the percentage of the AC sine wave that is applied (as indicated by the shaded portion). Because the waveforms have a sharp cut-off and cut-on characteristic when power is stopped and started over the course of a cycle, harmonic energy is produced. Sometimes lots of it.

Television sets and video cassette recorders (VCRs) are significant sources of EMI. There are three basic sources of EMI: vertical deflection circuits, horizontal deflection circuits, and the color circuits. The two deflection circuits use non-sine waveforms. The vertical operates at 59.94 Hz, while the horizontal operates at 15,734 Hz. The harmonics of both can cause problems with receivers and other devices, especially the horizontal deflection.

The color circuits of television sets use a phase reference at about 3.58 MHz. The EMI from the color phase circuits is especially strong from VCRs in the record mode. In my neighborhood, I can use my ham radio receiver to determine whether or not there is a popular movie on TV. The cacophony from the 3.58 MHz signals from all those VCRs tears up the 80-meter ham band.

Figures 15-3A and *15-3B* show the time series and spectrum plots of about six seconds worth of signal recorded at my home. I used a 10-turn, 6" diameter loop of wire connected to the shielded cable that goes to

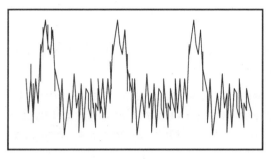

Figure 15-3A.
Noisy sine wave,
amplitude-vs-time.

the microphone input jack on a cassette recorder. The site was near my computer desk, so EMI from both the receiver and the monitor was recorded. The recorded signal was then processed in *Spectra Plus* software to produce the time and frequency series shown.

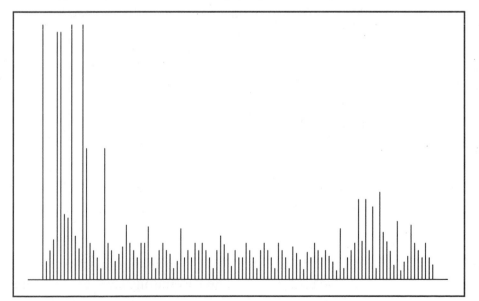

Figure 15-3B.
Noisy sine wave
spectrum.

Especially notice the spectrum in *Figure 15-3B*. This is a spectrum plot, so it shows lines representing frequencies, while the height of the line represents the strength of the signal. There is a huge concentration of energy around 60 Hz (left side of graph), with significant spikes at numerous frequencies out past 3,800 Hz. Because of the raucous noise on my Drake R-8A receiver, especially when it is tuned to the AM broadcast band, I suspect that significant harmonics at least go into the medium wave segment of the spectrum.

Tools for Finding EMI Sources

Rule 1: The correct tool for finding RF sources is anything that will permit you to unambiguously determine where the radiation is coming from. We will take a look at several low-cost possibilities.

If you are looking for a source that is out in the neighborhood somewhere, such as a loose power line or a malfunctioning electrical system in another building, then an ordinary solid-state portable AM BCB radio receiver may do the trick. Open the radio and locate the loopstick antenna. If you rotate the radio through an arc where the loopstick is first broadside to the arriving signal, and then perpendicular to the signal, you will notice a tremendous reduction in signal. Ferrite loopsticks are extremely sensitive to direction of arrival, with a sharp null occurring off the ends.

By noting the direction in which the null occurs (*Figure 15-4*), you find the line of direction to/from the signal source. It is not unambiguous, however. To find the actual direction from the two possibilities that are 180 degrees opposed, move along the line and note in which direction the signal increases. Except in a very few cases where reflections occur, the signal will become stronger as you move towards the source.

The standard medium wave and shortwave loop antenna used by a lot of radio enthusiasts is also useful for finding RF emitting sources. A square loop of 10 to 24 inches to the side is relatively easy to construct, and forms a very directional antenna. These antennas are commonly used in

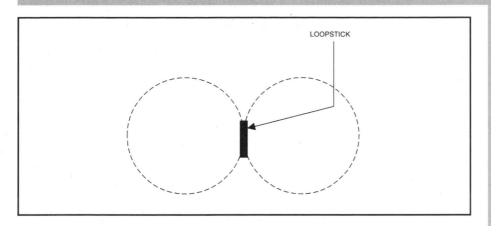

Figure 15-4.
Directional pattern
of a ferrite loopstick
antenna.

radio direction finding. Such loop antennas offer a null when pointed broadside to the signal direction, and a peak when orthogonal to the signal direction.

A portable radio with an S-meter will work wonders in this respect. An ad-hoc signal strength meter ("S-meter") can be formed using the circuit in *Figure 15-5*. This circuit plugs into the earphone jack of the receiver. The received audio is rectified by D1/D2 (a voltage doubler), and then applied to a DC current meter. Because resistor R1 (sensitivity control) is in series with M1, it operates as an analog voltmeter.

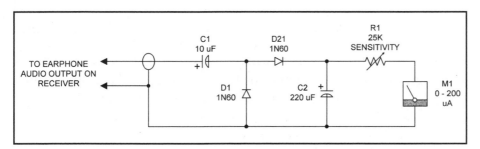

Figure 15-5.
"Audio S-meters" for
use on earphone-
equipped receivers.

If the noise peaks in the HF shortwave bands, then the radio's loopstick is of little use (it only works on the MW and LF AM BCBs). The HF antenna in those receivers is a telescoping whip. You can, however, make an external loopstick sensing antenna. A 7" ferrite rod is wound with about 20 turns of fine enameled wire. This wire is connected to a coaxial cable, that is in turn connected to the external antenna terminals of the receiver (if such exists). If there are no external antenna terminals, then a small coil of several turns wrapped around the telescoping antenna will couple signal to the radio. It's a good idea to keep the antenna at minimum height in order to minimize pickup from sources other than the loopstick. More sensitivity can be had if a resonant loopstick is available, but those also restrict the frequency band.

Two coil sensors are shown in *Figure 15-6*. The sensor in *Figure 15-6A* is a solenoid-wound inductor connected to a length of coaxial cable. The coil has a diameter of about 1 to 3 inches, and a length that is at least its own diameter. The coil consists of 2 to 10 turns of wire, depending on frequency (the higher the frequency, the fewer the turns required). The other end of the coaxial cable is connected to a receiver. When the coil is brought nearer the noise source, the signal level in the receiver will get higher.

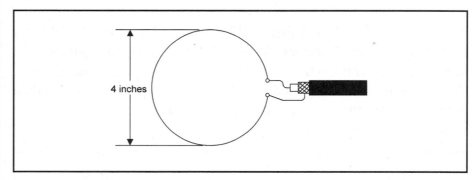

Figure 15-6A.
Single-turn loop
sniffer.

A single-turn "gimmick" is shown in *Figure 15-6B*. This sensor consists of a single loop, approximately 4" in diameter, connected to coaxial cable. It can be used well into the low end of the VHF spectrum, as well as at HF. The loop is made of either small-diameter copper tubing, heavy brass wire or rod, or heavy-gauge solid copper wire (#10 AWG or lower). The loop has some directivity, so can be used to ferret out extremely localized sources.

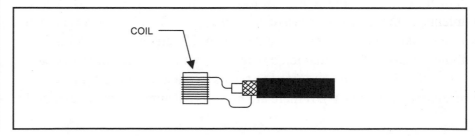

Figure 15-6B.
"Gimmick" coil
sniffer.

A fault with the single-turn loop is that it is somewhat sensitive to magnetic fields, although not so much as some of the other forms. *Figures 15-7A* and *15-7B* show dual-loop sensors that are less sensitive to magnetic field pickup. In *Figure 15-7A* the loops of the sensor are rectangular, and crossed in the center. The feedline (coaxial cable) is connected to a break in one of the two coils.

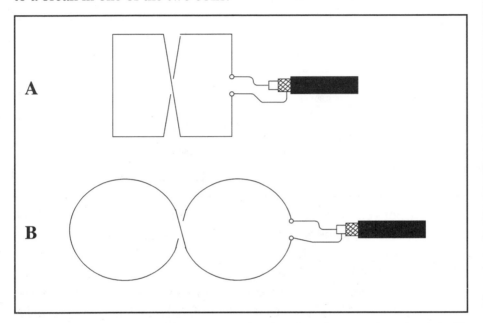

Figure 15-7.
Two forms of anti-magnetic field sniffer.

The version shown in *Figure 15-7B* is circular. One version that I built was made of #8 AWG solid copper wire formed around a tin can. Getting the coils about the right size and reasonably circular is relatively easy given a proper former. The exact size of the coils in *Figures 15-7A* and *15-7B* is not terribly important. The coil should not be too large, however, or it will be less local and may become a bit ambiguous in locating some sources.

RF Detectors

The RF output of the sensor coils can be routed to a receiver, and for low-level signals may well have to be so treated. For higher-power sources, however, an RF detector probe is used. *Figures 15-8* and *15-9* show two forms of suitable RF detector probe.

The RF detector in *Figure 15-8* is passive; i.e., it has no amplification. It can be used around transmitters and other RF power sources. The input from the sensor is applied to C1, a small value capacitor, and then is rectified by the 1N60 diode. The 1N60 is an old germanium type diode, and is used in preference to silicon diodes because it has a lower junction potential (so is more sensitive). The junction potential of Ge devices is 0.2 to 0.3 volts, and for silicon it is 0.6 to 0.7 volts. The pulsating DC from the rectifier is filtered by capacitor C2. Resistor R1 forms a load for the diode and is not optional.

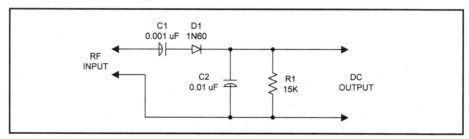

Figure 15-8.
RF detector converts
RF signal to DC.

An amplifier version of the RF detector circuit is shown in *Figure 15-9*. In this version the same RF detector circuit is used, but it is preceded by a 15 to 20 dB gain amplifier. In this particular circuit the amplifier is a Mini-Circuits MAR-1 device. It produces gain from near-DC to about 1,000 MHz. Other devices in the same series will work to 2,000 MHz, and in the related ERA-x series up to 8,000 MHz. Clearly, any of these devices is well suited to the needs of most readers. Cost of the MAR-1 device is very low (in the USA it is about $3 in unit quantities).

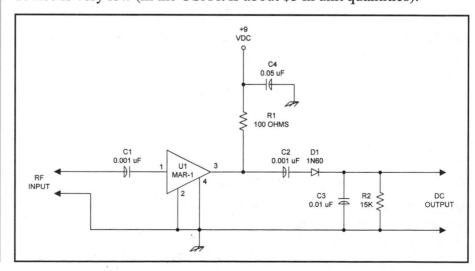

Figure 15-9.
Amplified RF
detector.

Dealing With AM and LW Broadcast Band EMI

In the late 1950s and early '60s I had two close friends who gained their Novice, and later General, class amateur radio licenses at about the same time as I did. One of them was John and the other was Doug. John lived across the street from WARL, a 1,000-watt country music station, and for awhile it rendered his *Hammarlund* HQ-110 receiver nearly unusable on the lower HF bands. The overload was that severe!

The location where John lived is known among radio buffs even today as "Intermod Hill" because of the huge amount of radio signals found there. It is one of the highest spots in Arlington County, VA*, so found natural use as a radio transmitter site. Today, WARL is now WABS at 5,000 watts. There are also the following stations sharing the site: WAVA-FM (50,000 watts—which shares the WABS tower), WETA-FM (50,000 watts), and the AT&T Long Lines Division microwave relay tower. In addition, the two broadcast towers (WABS/WAVA and WETA) each host a large number of two-way radio landmobile, cellular and other forms of communication antenna. The stations earn additional revenue by having height for hire to other radio users.

My friend who lived on Intermod Hill had to do two things to cohabitate with WARL. First, he had to open the HQ-110 and improve the shielding of the input circuitry. He fashioned a sheet metal screen, and changed the wire from the antenna jack to the input coils to coaxial cable. He also replaced the two-terminal antenna input that was found on early HQ-110s with a coaxial connector. The second thing he did was add a high-pass filter to the receiver antenna input external to the receiver. That filter method is the subject of this article.

Over the past several years I have written several articles, and no other topic in RadioScience Observing generated more mail than the articles on AM broadcast band interference. If you live anywhere near an AM BCB station, then you might have problems, even with a good receiver (although one distinguishing feature to justify the higher prices of good receivers is better performance in overload situations). It's obviously a "hot topic."

*Arlington County forms a perfect square, 10 miles to a side, with Washington, DC. The 10-mile square district was authorized in the US Constitution for the national capital, so was once part of the District of Columbia. It was ceded back to Virginia before the Civil War.

In the USA we use only the 530 to 1700-kHz medium wave AM BCB. Most stations operate with 1,000 to 50,000 watts ("clear channel" stations) of RF output power (although a few 250- to 500-watt local fizzlers also exist). Many stations either go off the air at sundown to protect distant stations on the same frequency, or either radically alter their antenna pattern, reduce power, or both at sundown. A few stations are designated "clear channel" stations, and operate with 50,000 watts, 24-hours a day. These stations (e.g., WSM Nashville, 650 kHz) are on frequencies that are not assigned to other stations for a distance of, I believe, 1300 miles radius. If you live within a few hundred meters of a station anything like those clear-channel blowtorches, then it's possible to see more than one volt of RF appearing at your receiver antenna terminals (one laboratory measured 4 volts in one case!).

In Europe, the receiver also has to contend with LW BCB stations, some of which are powerful enough to make our 50-gallon clear channel stations look more like cigarette lighters than blowtorches. Given that your receiver likes to see signals in the dozens of microvolts level, then you can understand the problem.

The Problem

So what is the problem? Your receiver, no matter what frequency it receives, is designed to accept only a certain maximum amount of radio frequency energy in the front-end. If more energy is present, then one or more of several *overload conditions* results. The overload could result from a desired station being tuned in too strong. In other cases, there are simply too many signals within the passband for the receiver front-end to accommodate. In still other cases, a strong out-of-band signal is present. *Figure 15-10* shows several conditions that your receiver might have to survive. *Figure 15-10A* is the ideal situation. Only one signal exists on the band and it is centered in the passband of the receiver. This never happens, and the problem has existed since Marconi was hawking interest in his wireless company. Indeed, Fessenden and Marconi interfered with each other while reporting yacht races off Long Island around the turn of the century. Not an auspicious beginning for maritime wireless! A more realistic situation is shown in *Figure 15-10B*. Here we see a

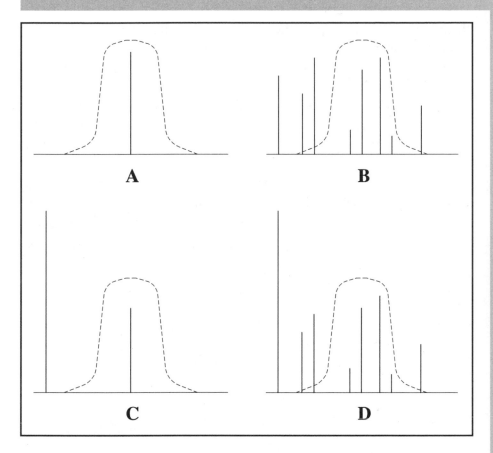

Figure 15-10.
Signal situations
confronting the front
end of your receiver:
A) ideal, one signal
inside the passband;
B) some signals
inside passband,
others outside;
C) extremely strong
local station outside
the passband, one
desired signal inside
passband;
D) mixture of strong
and weaker signals
in and out of
passband.

large number of signals—both in- and out-of-band signals—both weaker and stronger than the desired signal. Another situation is shown in *Figure 15-10C* where an extremely strong local station (e.g., AM BCB signal) is present, but is out of the receiver's front-end passband. The situation that you probably face is shown in *Figure 15-10D*: a large out-of-band AM BCB signal as well as the usual huge number of other signals both in and out of band.

Several different problems result from this situation, all of which are species of front-end overload-caused intermodulation and/or crossmodulation.

Blanketing. If you tune across the shortwave bands, especially those below 10 or 12 MHz, and note an AM BCB signal that seems like it is hundreds of kilohertz wide, then you are witnessing blanketing. It drives the mixer or RF amplifier of the receiver clean out of its mind, producing a huge number of spurious signals, and apparently a very wide bandwidth.

Desensitization. Your receiver can only accommodate a specified amount of energy in the front-end circuits. This level is expressed in the *dynamic range* specification of the receiver, and is hinted by the *third-order intercept point* (TOIP) and *-1 dB compression point* specifications. *Figure 15-11A* shows what happens in desensitization situations when a strong out-of-band signal is present. The strong out-of-band signal takes up so much of the dynamic range "headroom" that only a small amount of

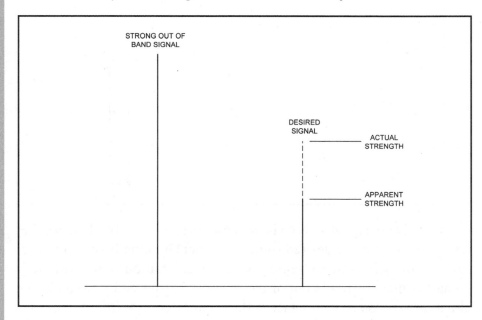

Figure 15-11A. Desensitization of receiver by strong local signal.

capacity remains for the desired signal. The signal level of the desired signal is thereby reduced to a smaller level. In some cases, the overload is so severe that the desired signal becomes inaudible. If you can filter out or otherwise attenuate the strong out-of-band signal (*Figure 15-11B*), then the headroom is restored, and the receiver has plenty of capacity to accommodate both signals.

Figure 15-12 shows two more situations. In *Figure 15-12A* we see the response of the receiver when output level is plotted as a function of

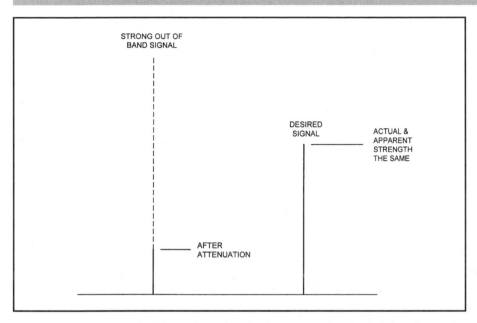

STRONG OUT OF
BAND SIGNAL

DESIRED
SIGNAL

ACTUAL &
APPARENT
STRENGTH
THE SAME

AFTER
ATTENUATION

Figure 15-11B.
Effect of filtering.

input signal level. The ideal situation is shown by the dotted line from the 0,0 intercept to an infinitely strong signal. Fat chance! Real radio receivers depart from the ideal and eventually saturate (solid line beyond the dot). The point denoted by the dot on the solid line is the point at which the TOIP is figured, but that's a topic for another time. What's important here is to consider what happens when signals are received that are stronger than the input signal that produces the flattening of the response.

In *Figure 15-12B* we see the generation of *harmonics*; i.e., integer (1, 2, 3,...) multiples of the offending signal's fundamental frequency. These harmonics may fall within the passband of your receiver, and are seen as valid signals even though they were generated in the receiver itself!

The intermodulation problem is shown in *Figure 15-12C*. It occurs when two or more signals are present at the same time. The strong intermodulation products are created when two of these signals heterodyne together. The heterodyne ("mixing") action occurs because the receiver front-end is nonlinear at this point. The frequencies produced by just two input frequencies (F1 and F2) are described by mF1 ± nF2, where m and n are integers. As you can see, depending on how many frequencies are present and how strong they are, a huge number of spurious signals can be generated by the receiver front-end.

So, what about IF selectivity? You have an IF filter of 2.7 to 8 kHz (depending on model and mode), so why doesn't it reject the dirty smelly bad-guy signals? The problem is that the damage occurs in the front-end section of the receiver, before the signals encounter the IF selectivity filters. The problem is due to an overdriven RF amplifier, mixer or both. The only way to deal with this problem is to reduce the level of the offending signal.

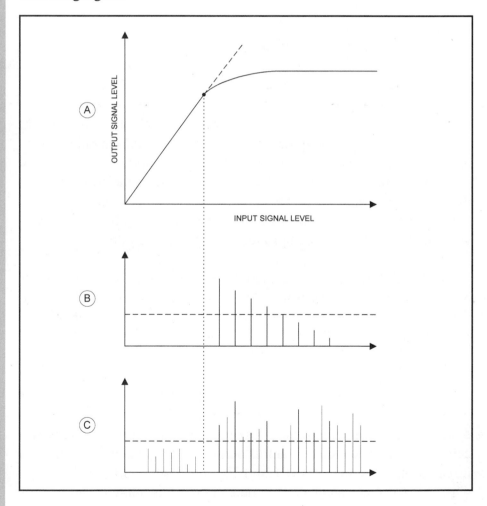

Figure 15-12. The source of the problem is overloading the front end of the receiver, creating signal products.

The Attenuator Solution

Some modern receivers are equipped with one or more switchable attenuators in the front-end. Some receivers also include an RF gain control that sometimes operates in the same manner. Some receiver operators use external switchable attenuators for exactly the same purpose. The idea behind the attenuator is to reduce all of the signals to the front-end enough to drop the overall energy in the circuit to below the level that can be accommodated without either overload or intermodulation occurring at significant levels. The attenuator reduces both desired and undesired signals, but the perceived ratio is altered when the receiver front-end is de-loaded to a point where desensitization occurs, or intermods and harmonics pop up.

The Antenna Solution

The antenna you select can make some difference in AM BCB problems. Generally, a resonant HF antenna with its end nulls pointed toward the offending station will provide marginally better performance than a random length wire antenna (which are popular amongst SWLs). Also, it is well known that vertical HF antennas are more susceptible to AM BCB because they respond better to the ground-wave electrical field generated by the BCB station.

The Filter Solution

One of the best solutions is to filter out the offending signals before they hit the receiver front-end, while affecting the desired signals minimally. This task is not possible with the attenuator solution, which is an "equal opportunity" situation because it affects all signals equally. *Figure 15-13A* shows what happens to a signal that is outside the passband of a frequency selective filter: it is severely attenuated. It does not drop to zero, but the reduction can be quite profound in some designs.

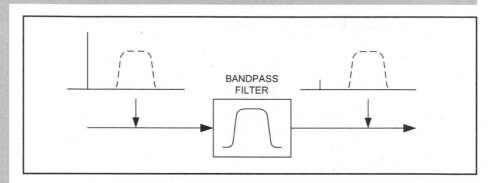

Figure 15-13A.
Effect of filter on
out-of-band signals
is a large insertion
loss.

Signals within the receiver's passband are not unaffected by the filter, as shown in *Figure 15-13B*. The loss for in-band signals is, however, considerably less than for out-of-band signals. This loss is called *insertion loss*, and is usually quite small (1 or 2 dB) compared to the loss for out-of-band signals (lots of dB).

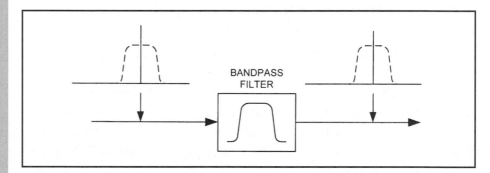

Figure 15-13B.
Effect of filter on in-band signals is a
small insertion loss.

Several different types of filter are used in reducing interference. A *high-pass filter* passes all signals *above* a specified cutoff frequency (F_c). The *low-pass filter* passes all signals *below* the cutoff frequency. This filter is the type that hams using HF transmitters place between the transmitter and antenna to prevent harmonic radiation from interfering with television operation. A *bandpass filter* passes signals between a lower cutoff frequency (F_L) and an upper cutoff frequency (F_H). A *stop-band* filter is just the opposite of a bandpass filter: it stops signals on frequencies between F_L and F_H, while passing all others. A *notch* filter, also called a wave trap, will stop a particular frequency (F_o), but not a wide band of frequencies as does the stop-band filter. In all cases, these filters stop the frequencies in the designated band, while passing all others...more or less.

The positioning of the filter in your antenna system is shown in *Figure 15-14*. The ideal location is as close as possible to the antenna input connector. The best practice, if you have the space at your operating position, is to use a double-male coaxial connector to connect the filter output connector to the antenna connector on the receiver. A short piece of coaxial cable can connect the two terminals if this approach is not suitable in your case. Be sure to earth both the ground terminal on the receiver and the ground terminal of the filter (if one is provided). Otherwise, depend on the coaxial connectors' outer shell making the ground connection.

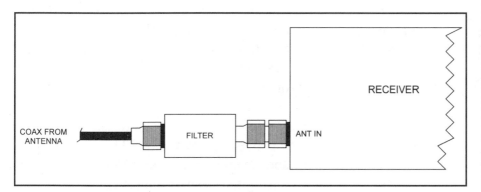

Figure 15-14. Connection of the filter to the receiver.

Wave traps. A wave trap is a tuned circuit that causes a specific frequency to be rejected. Two forms are used: *series tuned* (*Figure 15-15A*) and *parallel tuned* (*Figure 15-15B*). The series tuned version is placed across the signal line (as in *Figure 15-15A*), and works because it produces a very low impedance at its resonant frequency and a high impedance at frequencies removed from resonance. As a result, the interfering signal will see a resonant series-tuned wave trap as a short circuit, while other frequencies do not. The parallel resonant form is placed in series with the antenna line (as in *Figure 15-15B*). It provides a high impedance to its resonant frequency, so will

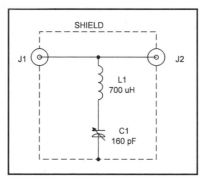

Figure 15-15A. Series-tuned wave trap.

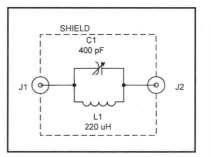

Figure 15-15B. Parallel-tuned wave trap.

block the offending signal before it reaches the receiver. It provides a low impedance to frequencies removed from resonance.

The wave traps are useful in situations where a single station is causing a problem, and you don't want to eliminate nearby stations. For example, if you live close to an MW AM BCB signal and don't want to interrupt reception of other MW AM BCB signals or LW AM BCB signals. The values of components shown in *Figures 15-15A* and *15-15B* are suitable for the MW AM BCB, but can be scaled to the LW BCB if desired.

If there are two stations causing significant interference, then two wave traps will have to be provided, separated by a short piece of coaxial cable. In that case, use a parallel tuned wave trap for one frequency, and a series tuned wave trap for the other. Otherwise, interaction between the wave traps will cause problems.

High-Pass Filters. One very significant solution is to use a high-pass filter with a cutoff frequency between 1700 and 3000 kHz. It will pass the shortwave frequencies, and severely attenuate AM BCB signals in both MW and LW bands, causing the desired improvement in perfor-mance. *Figure 15-16* shows a design used for many decades. It is easily built because the capacitor values are 0.001 µF and 0.002 µF (which some people make by parallel connecting two 0.001 µF capacitors). The inductors are both 3.3 µH, so can be made with toroid cores. If the T-50-2 RED cores are used (A_L = 49), then 26 turns of small-diameter enameled wire will suffice. Or if the T-50-15 RED/WHITE cores are used (A_L = 135), then 15 turns are used. The circuit of *Figure 15-16* produces pretty decent results for low effort.

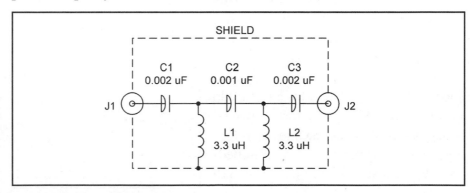

Figure 15-16. Standard AM BCB protection for HF receivers is this high-pass filter.

Absorptive Filters. The *absorptive filter* (Orr 1996 and Weinreich/Carroll 1968) solves a problem with the straight high-pass filter method, and produces generally better results at the cost of more complexity. This filter (*Figure 15-17*) consists of a high-pass filter (C4-C6/L4-L6) between the antenna input (J1) and the receiver output (J2). It passes signals above 3 MHz and rejects those below that cutoff frequency. It also has a low-pass filter (C1-C3/L1-L3) that passes signals below 3 MHz. What is notable about this filter, and from which it takes its name, is the fact that the low-pass filter is terminated in a 50-ohm dummy load. This arrangement works a little better than the straight high-pass filter method because it absorbs energy from the rejected band, and reduces (although does not eliminate) the effects of improper filter termination.

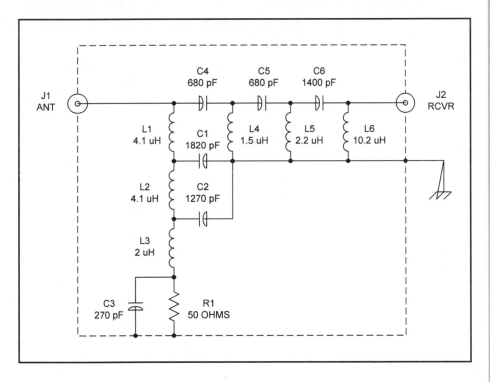

Figure 15-17.
Absorptive filter.

Some of the capacitor values are nonstandard, but can be made using standard disk ceramic or mica capacitors using combinations in *Table 15-1*:

Table 15-1.
Capacitor values.

C1:	1,820 pF.	Use 1000 pF (0.001 µF or 1 nF) in parallel with 820 pF
C2:	1,270 pF.	Use 1000 pF and 270 pF in parallel
C6:	1,400 pF.	Use 1000 pF, 180 pF and 220 pF in parallel.

The other capacitors are standard values.

The coils are a bit more difficult to obtain. Although it is possible to use slug-tuned coils obtained from commercial sources (e.g., Toko), or homebrewed, this is not the preferred practice. Adjusting this type of filter without a sweep generator might prove daunting due to interactions between the sections. A better approach is to use toroid core homebrew inductors. The toroidal cores reduce interaction between the coil's magnetic fields, so simplifies construction. Possible alternatives are shown below in *Table 15-2*:

Table 15-2.
Coil values.

Coil	Value	Core	A_L Value	Turns
L1	4.1 µH	T-50-15 RED/WHITE	135	17
L2	4.1 µH	T-50-15 RED/WHITE	135	17
L3	2 µH	T-50-15 RED/WHITE	135	12
L4	1.5 µH	T-50-2 RED	49	18
		T-50-6 YEL	40	20
L5	2.2 µH	T-50-2 RED	49	21
		T-50-6 YEL	40	24
L6	10.2 µH	T-50-2 RED	49	46
		T-50-6 YEL	40	51

For all coils use wire of a similar UK SWG size #24 to #30 AWG enamel insulation.

The dummy load used at the output of the low-pass filter (R1 in *Figure 15-17*) can be made using a 51-ohm carbon or metal-film resistor, or two

100-ohm resistors in parallel. In a pinch a 47-ohm resistor could also be used, but is not preferred. In any event, use only noninductive resistors such as carbon composition or metal-film 1/4 to 2-watt resistors.

If you would like to experiment with absorptive filters at other cutoff frequencies than 3 MHz, then use the reactance values in *Table 15-3* to calculate component values:

Component	X (X$_L$ or X$_C$)
L1	28.8Ω
L2	78.4Ω
L3	38Ω
L4	28.8Ω
L5	42Ω
L6	193Ω
C1	28.8Ω
C2	42Ω
C3	193Ω
C4	78.4Ω
C5	78.4Ω
C6	38Ω

Table 15-3.
Reactance values.

The exact component values can be found from variations on the standard inductive and capacitive reactance equations:

$$L_{\mu H} \ = \ \frac{X_L}{2 \pi F_c} \times 10^6 \quad \text{microhenrys} \qquad (15\text{-}1)$$

and,

$$C_{pF} \ = \ \frac{10^{12}}{2 \pi F_c X_c} \quad \text{picofarads} \qquad (15\text{-}2)$$

These component values are bound to be nonstandard, but can be made either using coil forms (for inductors) or series-parallel combinations of standard-value capacitors.

Stubs

At higher frequencies, where wavelengths are short, a special form of trap can be used to eliminate interfering signals. The half wavelength shorted stub is shown in *Figure 15-18*. It acts like a series resonant wave trap, so is installed in a similar manner right at the receiver. The stub is shorted at the free end, so presents a low impedance at its resonant frequency and a high impedance at all other frequencies. The length of the stub is given by:

$$L_{CM} = \frac{1250\ VF}{F_{MHZ}} \tag{15-3}$$

Where:

L_{CM} is the length in centimeters (cm)

F_{MHZ} is the frequency of the interfering signal in megahertz (MHz)

VF is the *velocity factor* of the coaxial cable (Typically 0.66 for polyethylene cable, and 0.80 for polyfoam. For other types see manufacturer data.)

Shielding

Shielding is a nonnegotiable requirement of filters used for the EMI reduction task. Otherwise, signal will enter the filter at its output and will not be attenuated. Use an aluminum shield box of the sort that has at least 5-6 mm of overlap of a tight fitting flange between upper and lower portions. I used a tinned steel RF box for this purpose when building the prototypes for the filters shown earlier.

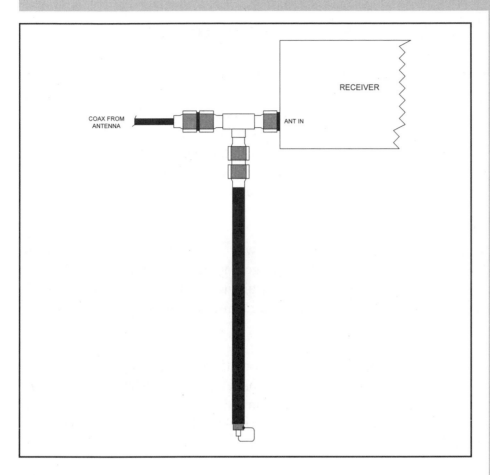

Figure 15-18.
Shorted stub to
form notch filter.

Expected Results

If the correct components are selected, and good layout practice is followed (which means separating input and output ends, as well as shielding the low-pass and high-pass sections separately), then the absorptive
filter can offer stopband attenuation of -20 dB at one octave above F_c,
-40+ dB at two octaves and -60 dB at three octaves. For a 3 MHz signal,
one octave is 6 MHz, two octaves are 12 MHz and three octaves are
24 MHz. My results were slightly less than these figures because some
of my components were ill-matched (e.g., slug-tuned commercial inductors were used rather than toroid core coils), but the results were consistent with expectations.

Dealing With 60 Hz Interference

Over the past several decades I've seen a lot of situations where 60 Hz interference disrupted all manner of circuits and instruments. It's basically a pain. The best solution, by far, is to prevent it from occurring. The usual methods involve shielding, or using a location that is free of 60 Hz fields. Lots of luck! Have you noticed those lights you are using to read this article? Unless you are on battery power, they are powered by 60 Hz alternating current (AC), or if in certain other countries than USA and Canada, 50 Hz AC.

A practical example arose when I was employed as an electronics technician (later as an engineer after I got my degree) at George Washington University Medical Center in Washington, D.C. A physiologist was using a Wheatstone bridge transducer to measure the contractions of some muscle or another.

The output of the sensor was very tiny. Its sensitivity factor was 5 mmV/V/gram-force. With a 6V battery, and gram-forces between 0.5 and 2, the output of the contraction pulses were from 15 to 60 µV. The bridge had a differential output, so had to be processed in a differential amplifier.

Figure 15-19 shows the system used. It is not unlike circuits used by a number of scientific instruments, many of which find use in radioscience. The two sensor output wires were passed to a very long (about 12 feet) shielded two-wire cable, to a differential input on a strip-chart recorder (shown here as a differential amplifier. Unfortunately, the differential input was not truly differential. It was—like an oscilloscope—a two-channel device ("A" and "B" channels) that permitted an "A-B" input to function as pseudo-differential).

Figure 15-19.
Poor instrumentation
scheme is susceptible
to noise pick-up.

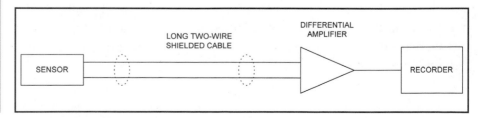

Even though the cable was shielded, the slight difference in the input amplifiers produced a common-mode voltage that was seen as a valid input signal by the recorder. The 60 Hz fields radiated by the building power wiring induced 60 Hz signals into the sensor circuit. And with the tiny input signals, any 60 Hz at all was fatal. The scientist's tracings were corrupted with 60 Hz interference.

My first solution was to tell him to buy a differential amplifier plug-in for the recorder. Wrong answer! It seems that research grants for junior scientists are not all that generous. He simply didn't have the money to buy the correct amplifier. So I decided to build a differential amplifier at the sensor end of the cable. *Figure 15-20* shows the resultant system. A gain-of-100 DC differential amplifier, using a CA-3240 operational amplifier, was built and installed right at the sensor. A single-conductor coaxial cable was run to one of the single-ended amplifier inputs of the recorder (eliminating the A-B pseudo-differential problem). The 60 Hz completely disappeared.

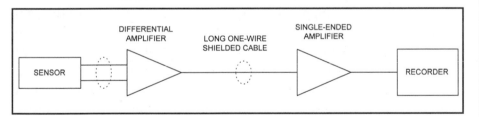

Figure 15-20. Using a differential amplifier at head-end of transmission line reduces interference.

Unfortunately, it's not always possible to prevent 60 Hz from occurring. In that event, we need to find a solution involving filtering out the 60 Hz stuff. If the input signal permits, then a high-pass or low-pass filter can be used. Any signal has harmonics, and these must fall within the bandpass of the filter. For very slow signals, with a maximum frequency content of 20 Hz or less, then a low-pass filter will sufficiently attenuate the 60 Hz signals. Similarly, if the signal frequencies are much higher than 60 Hz, then a high-pass unit will be needed.

Notch Filters

The usual solution to unwanted in-band frequencies is the *notch filter*. The frequency response of a notch filter is shown in *Figure 15-21*. These filters are similar to another class, *bandstop filters*, but the band of rejection is very narrow around the center frequency (F_c). The bandwidth (BW) of these filters is the difference between the frequencies at the two -6 dB points, when the out-of-notch response is the reference 0 dB point. These frequencies are F_L and F_H, so the bandwidth is $F_H - F_L$.

The "sharpness" of the notch filter is a measure of the narrowness of the bandwidth, and is specified by the "Q" of the filter. The Q is defined as the ratio of the center frequency F_c to bandwidth:

$$Q = \frac{F_c}{BW} \qquad\qquad (15\text{-}4)$$

For example, a notch filter that is centered on 60 Hz, and has -6 dB points of 58 and 62 Hz (4 Hz bandwidth) has a Q of 60/4 or 15.

The notch filter doesn't remove the entire offending signal, but rather suppresses it by a large amount. *Notch depth* (see again *Figure 15-21*) defines the degree of suppression, and is defined by the ratio of the gain of the circuit at an out-of-notch frequency (e.g., F_{ob}) to the gain at the notch frequency. Assuming equal input signal levels at both frequencies (which has to be checked—most signal generators have variable output levels with changes of frequency!), the notch depth can be calculated from the output voltages of the filter at the two different frequencies:

$$Notch\ Depth = 20\ LOG \qquad\qquad (15\text{-}5)$$

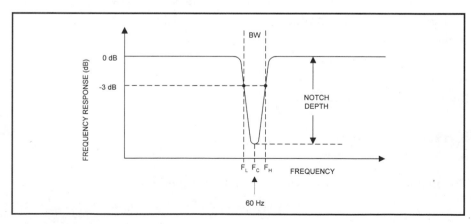

Figure 15-21.
Notch filter
response curve.

Twin-T Notch Filter Networks

One of the most popular forms of notch filter is the *twin-tee filter network*, shown in *Figure 15-22*. It consists of two T-networks, consisting of C1/C3/R2 and R1/R3/C2. Notch depths of -30 to -50 dB are relatively easy to obtain with the twin-tee, assuming proper circuit design and component selection. Very good matching and selection of parts makes it possible to achieve -60 dB suppression.

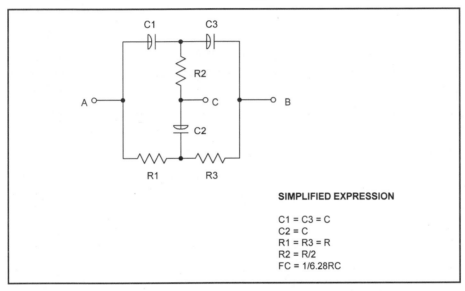

SIMPLIFIED EXPRESSION

C1 = C3 = C
C2 = C
R1 = R3 = R
R2 = R/2
FC = 1/6.28RC

Figure 15-22.
Twin-tee passive
R-C notch filter.

The center notch frequency of the network in the generic case is given by:

$$F_c = \frac{1}{2\pi} \sqrt{\frac{C1 + C3}{C1\ C2\ C3\ R1\ R3}} \qquad (15\text{-}6)$$

We can simplify the expression above by adopting a convention that calls for the following relationships:

C1 = C3 = C

R1 = R3 = R

C2 = 2C

R2 = R/2

If this convention is adopted, then we can reduce the frequency equation to:

$$F_c = \frac{1}{2 \pi R C}$$ (15-7)

In these expressions, F is in hertz (Hz), R is in ohms and C is in farads. Be sure to use the right units when working these problems: "10 kohms" is 10,000 ohms, and "0.001 µF" is 1×10^{-9} farads. In calculating values, it is usually prudent to select a capacitor value, and then calculate the resistance needed. This is done for two reasons: 1) there are many more standard resistance values; and 2) potentiometers can be easily used to trim the values of resistances, but it is more difficult to use trimmer capacitors. For 60 Hz filters, some common values for R and C are:

C	R
0.001 µF	2,652,582 Ω
0.01 µF	5,258 Ω
0.15 µF	17,684 Ω

One of the problems of these filters is that the depth of the notch is a function of two factors involving these components. First, that they are very close to the calculated values, and second, that they be matched closely together. For example, a 60 Hz notch filter was built using the 0.15 µF and 17,684 Ω values from the table above, the 0.15 µF capacitors were selected at random from a group of a dozen or so "mine run" capacitors of good quality, while the resistors were 18 kohm, 5% metal-film resistors. The notch depth at 60 Hz was only 10 dB, but at 58 Hz it was 48 dB. The mismatch caused a significant shift of notch frequency.

A second filter was built using the same values. In this case, the 0.15 µF capacitors were selected from about 20 on hand (precision components are difficult to obtain). In order to match them as close as possible, the capacitance of each was measured using the capacitance tester function on a low-cost (< $100) digital multimeter. The order of prior selection was to find those that closely matched *each other*, and only incidentally how close they come to the calculated value. Errors in the mean capacitance of the selected group can be trimmed out using a potentiometer in the resistor elements of the twin-tee network.

When selecting a frequency source, either select a well-calibrated source, or use a frequency counter to measure the frequency. Keep in mind the situation described above where only a 2 Hz shift produced a 38 dB difference in notch depth! Alternatively, use a 6.3-volt or 12.6-volt AC filament transformer secondary as the signal source. (WARNING: The primary circuits of these transformers are at a potential of 115 VAC, and can thus be lethal if mishandled!)

Adjustable-frequency notch filters can be built using the twin-tee idea, but none of the usual solutions are really acceptable. One implementation requires three ganged *matched* potentiometers or three ganged capacitors. Unfortunately, in either case at least one of the variable sections must be of different value from the other two, causing a tracking problem. You might not notice a tracking problem in some circuits, but in a high-Q notch filter it can be disastrous.

Active Twin-Tee Notch Filters

Active frequency selective filters use an active device such as an operational amplifier to implement the filter. In the active filter circuits to follow, the "twin-tee" networks are shown as block diagrams for sake of simplicity, and are identical to those circuits shown earlier; the ports "A," "B," and "C" in the following circuit are the same as in the previous network.

The simplest case of a twin-tee filter is to simply use it "as is," i.e., use the filter circuits shown above. But the better solution is to include the twin-tee filter in conjunction with one or more operational amplifiers. There is one solution in which the twin-tee network is cascaded with an input buffer amplifier (optional) and an output buffer amplifier (required). These amplifiers tend to be noninverting op-amp follower circuits. The purpose of buffer amplifiers is to isolate the network from the outside world. For low-frequency applications, the op-amps can be 741, 1458 and other similar devices. For higher-frequency applications, such as those with an upper cutoff frequency above 3 kHz, use a non-frequency compensated device such as the CA-3130 or CA-3140 devices.

A superior circuit is shown in *Figure 15-23*. In this circuit, port-C of the twin-tee network (the common point) is connected to the output terminal of the output buffer amplifier. There is also a feedback network consisting of two resistors (R$_a$) and a capacitor (C$_a$). The values of R and C in the twin-tee network are found from the equation above, while the values of R$_a$ and C$_a$ are found from:

$$R_a = 2\,R\,Q \qquad\qquad (15\text{-}8)$$

and,

$$C_a = \frac{C}{Q} \qquad\qquad (15\text{-}9)$$

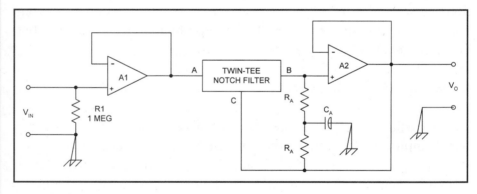

Figure 15-23.
Op-amp buffered
notch filter.

Example

Design a 60 Hz notch filter with a Q of 8.

    1. Select a trial value for C: 0.01 µF.

    2. Calculate the value of R from the equation: 265,392 Ω

    3. Calculate R/2: 265,392/2 = 132,696 Ω

    4. C2 = 2C = (2)(0.01 µF) = 0.02 µF.

    5. Select R$_a$: R$_a$ = 2QR = (2)(8)(265,392 Ω) = 4.24 megΩ

    6. Select C$_a$ = C/Q = 0.01 µF /8 = 0.0013 µF.

When *Figure 15-23* was built using these values in *Joe's Basement Lab*, using the twin-tee network the null was close to -48 dB deep using components at hand.

A variable Q control is shown in *Figure 15-24*. In this circuit, a noninverting follower (A3) is connected in the feedback loop in place of R_a and C_a. The Q of the notch is set by the position of the 10 kohm potentiometer (R2). Values of Q from 1 to 50 are available from this circuit.

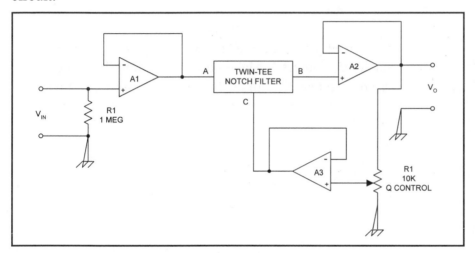

Figure 15-24.
Variable notch filter using potentiometers for Q control.

Another approach to notch filter circuits is shown in *Figure 15-25*. This circuit is sometimes called the *gyrator* or *active inductor* notch filter (it's also sometimes called the *virtual inductor notch filter*).

The notch frequency is set by:

$$F_c = \frac{1}{2\pi\sqrt{R_a R_b C_a C_b}} \tag{15-10}$$

Equation 15-10 can be simplified to...

$$F_c = \frac{1}{2\pi R\sqrt{C_a C_b}} \tag{15-11}$$

...if the following conditions are met:

$$\frac{R3}{R1} = \frac{R2}{R_a + R_b} = \frac{R2}{2R}$$

It is possible to use any one of the elements, C_a, C_b, R_a, or R_b, to tune the filter. In most cases, C_a is made variable and C_b is a large-value fixed capacitor. The 1,500 pF variable capacitor can be made by paralleling all sections of a three-section broadcast variable, with a single small fixed or trimmer capacitor. Alternatively, since most applications will require a trimmer rather than a big honkin' broadcast variable, it is also possible to parallel one or more small capacitors and a trimmer. For example, a 100 pF trimmer can be paralleled with a 1,000 pF and 470 pF to form the 1,500 pF capacitance required. Make sure to use low-drift, precision capacitors. Or you can match them using a digital capacitance meter.

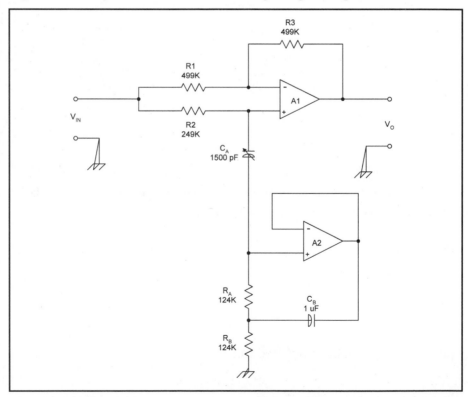

Figure 15-25.
Gyrator notch filter.

Be careful when using any filter to remove components from a wave-form. If the filter is not a high-Q type, then too much of the signal might be removed. In medical electrocardiograph (ECG) systems the signal has components from 0.05 to 100 Hz, so 60 Hz is right in the center of the range! Oops. To make matters worse, the leads have to be connected to the human body, so are unshielded at their very ends. Interference from 60 Hz is almost guaranteed unless care is taken. But filtering can take out components that assist the physician in making diagnosis, so is only used when it is unavoidable. On medical ECG amplifiers the filter is usually switchable so it can be either in or out of the circuit.

Noise Canceling Bridge

Figure 15-26 shows a circuit that can be used on receivers. Two antennas are needed. J1 is connected to the main antenna used with the receiver, while J2 is connected to a small whip antenna, the sort used on portable shortwave and FM radio receivers. The output (J3) is connected

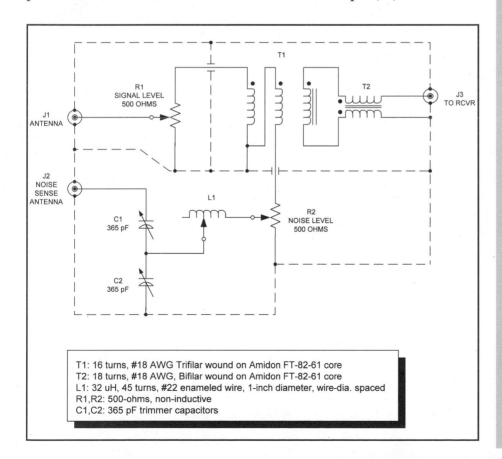

T1: 16 turns, #18 AWG Trifilar wound on Amidon FT-82-61 core
T2: 18 turns, #18 AWG, Bifilar wound on Amidon FT-82-61 core
L1: 32 uH, 45 turns, #22 enameled wire, 1-inch diameter, wire-dia. spaced
R1,R2: 500-ohms, non-inductive
C1,C2: 365 pF trimmer capacitors

Figure 15-26.
Noise eliminator for
HF receivers.

to the antenna input of the receiver. The bridge works by picking up RF harmonics of the 60 Hz power line, which can afflict reception from the VLF through MW bands, phase shifting them 180 degrees, so that they will cancel the noise pulses arriving on the main antenna. Because the whip antenna is much less sensitive to the signals being sought, they are affected only a small amount (insertion loss). Adjusting this circuit can be tricky, but it is possible to make a substantial improvement.

Chapter 16
SOME INSTRUMENT DESIGN RULES

Chapter 16
Some Instrument Design Rules

One of the aspects of RadioScience Observing that attracts many amateur scientists is the fact that they must build some of their own instrumentation (unless they are rich, or lucky in the surplus market). Even in schools and colleges, budgets are rarely so generous as to permit unlimited instrument buying. Some homebrewing, adaptation, improvising and out and out chicanery is sometimes needed.

The process of designing electronic instrumentation circuits or projects is not an arcane art, open only to a few highly skilled initiates. Rather, it is a logical step-by-step process that can be learned. Like any skill, design skill is improved with practice, so one is cautioned against both excessive expectations the "first time at bat" and discouragement if the process did not exactly work out as planned the first time.

Some of the material in this chapter may be called "philosophy," and that may be a fair label to attach. Although many technical people claim to disdain "philosophy," we all have it and use it. It's just that some people think about it a lot, others think about it either a little or at only a few times, while others don't think about it at all—they take actions based not on a logically considered viewpoint, but rather, by dumb luck and pure default.

A Design Procedure

The procedure that one adopts to designing may well be different from what is presented below, and that's alright. The purpose of offering a procedure is to *systematize* the process. While it is conceivable that one can design an instrument by a process similar to Brownian motion, it is the systematic approach that most often yields success. This procedure assumes a product that is a one-of-a-kind instrument, as in a scientific laboratory or plant. Designing a product for production and sale follows a similar procedure, but involves marketing and production problems as well. The steps in the procedure, some of which are iterative with respect to each other, are offered below:

1. Define and tentatively solve the problem. Create multiple tentative solutions wherever feasible.

2. Determine the critical attributes required of the final product; incorporate these into a specification and a test plan that determines objective criteria for acceptance or rejection.

3. Determine the critical parameters and requirements.

4. Attempt a "block diagram" solution.

5. Apportion requirements (e.g., gain, frequency response, etc.) to the various blocks.

6. Make a first-cut guess at which functions ought to be in software (if appropriate) and which in hardware.

7. Perform analysis and do simulations on the block diagram to test validity of the approach.

8. Design specific circuits to fill in the blocks.

9. Build and test the circuits, and write the software.

10. Combine the circuits with each other on a breadboard.

11. Test the breadboarded circuit according to a fixed test plan.

12. Build brassboard that incorporates all changes made in the previous steps.

13. Test brassboard and correct problems.

14. Integrate software and hardware.

15. Design and construct final configuration.

16. Test final configuration.

17. Ship the product.

Solving the (Right) Problem

The purpose of the designer is to solve some problem or another using analog circuits, digital circuits, a computer or whatever else is available in the armamentarium. There are two related problems often seen in the efforts of novices.

First, it is often the case that the designer will have a tentative favorite approach in mind before the problem is properly understood. Decisions are made based on what the designer is most comfortable doing. For example, many younger designers are likely to select the digital solution in a knee-jerk manner that excludes any consideration of the analog solution. Both should be evaluated, and the one that best fits the need selected.

Second, *be sure you understand the problem being solved*. While this advice seems trivial, it is also true that failure to understand the problem at hand often sinks designs before they have a chance to manifest themselves. There are several facets to this problem. For example, a natural tendency of engineers is to think that an elegant solution is complex and large-scale. If this mistake is made, then it is likely that the product will be overdesigned and have too many whistles and bells. It was, after all, designed to solve a much harder problem than was actually presented.

Another aspect to understanding the problem is to understand the final customer's *use* for the product. It is all too easy to get caught up in the specification, or our own ideas about the job, and overlook altogether what the user needs to accomplish with the product. An example is derived from biomedical instrumentation. A physiologist requested a pressure amplifier that would measure blood pressures over a range of 0 to 300 mmHg (Torr). What the salesman never told the plant was: 1) it would be used on humans (safety and regulatory issues); 2) that blood would come in contact with the diaphragm (cleaning and/or liquid isolation issues); and 3) it would occasionally be used for measuring 1-to-5 mmHg central venous pressures (which implies low-end linearity issues).

Part of the problem in determining the level of complexity, or the specific design's function, is miscommunication between the end user and the designer. Although miscommunications occur frequently between in-house designers and their "clients," it is probably most common between distant customers and engineers in the plant. Of course, marketing people may never let the engineer and customer get together (either from igno-rance or a fear that the salesman's little lies will surface: "The reason I hate engineers is that, under duress, they tend to blurt out the truth.").

The proper role of the designer is to scope out the problem at hand, understand what the circuit or instrument is supposed to do, how and where the user is going to use it, and exactly what the user wants and expects from the product.

Determine Critical Attributes

This step is basically the fruit of understanding and solving the correct problem. From the solution of the problem one can determine, and write down, a set of attributes, characteristics, parameters and other indices of the product's final nature.

It is at this point that one must write a specification that documents what the final product is supposed to do. The specification must be clearly written so that others can understand it. A concept or idea does not really exist, except in the mind of the originator. One must, according to W. Edwards Deming[*], create *operational definitions* for the attributes of the product. One cannot simply say "it must measure pressure to a linearity of 1 percent over a range of 0 to 100 p.s.i." Rather, it might be necessary instead to specify a test method under which this requirement can be met. There might be, after all, more than one standard for pres-sure and measurement, and there is certainly more than one definition of linearity. The operational definition serves the powerful function of pro-viding everyone with the same set of rules—basically, it levels the play-ing field.

[*] *Out of the Crisis*, W. Edwards Deming, MIT Press.

Part of this step, and of making an operational definition, is to write a test plan for the final product. It is here that one determines (and often contractually agrees) exactly what the final product will do, and defines the objective criteria of goodness or badness that will be used to judge the product.

Determine Critical Parameters and Requirements

Once the product is properly scoped out, it is time to determine the critical technical parameters that are needed to meet the test requirements (and hopefully the user's needs—if the test requirements are properly written). Parameters such as frequency response, gain and so forth tend to vary in multimode instruments, so one must determine the worst case for each specification item and design for it.

Attempt a Block Diagram Solution

The block diagram is a signal flow or function diagram that represents stages, or collections of stages, in the final instrument. In large instruments there might be several indexed levels of block diagram, each one going into finer detail.

Apportion Requirements To The Blocks

Once the block diagram solution is on paper, tentatively apportion system requirements to each block. Distribute gain, frequency response, and other attributes to each block. Keep in mind that factors such as gain distribution can have a profound effect on dynamic range. Also, the noise factor and drift of any one amplifier can have a tremendous effect on the final performance—and that it's in these types of parameters (where critical placement of one high-quality stage may be sufficient) that added cost and complexity often arises.

Analyze and Simulate

Once the block diagram is set, and the requirements apportioned to the various stages, it is time to analyze the circuit and run simulations to see if it will actually work. A little "desk checking" goes a long way towards eliminating problems later on when the thing is first prototyped. Plug in typical input values and see what happens on a stage-by-stage basis. Check for the reasonableness of outputs at each stage. For example, if the input signal should drive an output signal to, say, 17 volts, and the operational amplifiers are only operated from 5-volt power supplies, then something is wrong and will have to be corrected.

Design Specific Circuits For Each Block

It is at this point that the remainder of this book is of most usefulness to you, for it is on the specific circuits that we will dwell later. In this step, fill in those blocks with real circuit diagrams.

Built and Test The Circuits

At this point one must actually construct the individual circuits, and test them to make sure they work as advertised (unless, of course, the circuit is so familiar that no testing is needed). Keep in mind that some of your best ideas for simplified circuits may not actually work—and this is the place to find out. Use a benchtop breadboard that allows circuit construction using plug-in stripped-end wires.

Combine the Circuits in a Formal Breadboard

Once the validity of the individual circuits is determined, combine them together in a formal breadboard. Whether built on a benchtop breadboard (as above), or on a prototyping board, make sure that the layout is similar to that expected in the final form.

Test the Breadboard

Test the overall circuit according to a formally established objective criteria. This test plan should be developed earlier (step 2). As problems arise and are solved, make changes and/or corrections—and *document the results*. It is, perhaps, the main failing of inexperienced designers that they do not properly document their work, even in an engineer's or scientist's notebook.

Build and Test a Brassboard Version

A "brassboard" is a version made as close as possible to the final configuration. While breadboarding techniques can be a little sloppy, the brassboard should be a proper printed-circuit board. The test criteria should be the same as before, updated only for changes that occurred. If problems turn up, they should be corrected prior to proceeding further. Keep in mind that the most common problems that occur in leaping from breadboard to brassboard are layout (e.g., coupling between stages), power distribution, and ground noise (these are the areas of difference between the two configurations).

Design, Build and Test Final Version

Once all of the problems are known, solved and changes incorporated, it is time to build the final product as it will be given to the end user. It is at this point that the reputation of the designer is made or broken, because it is here that the product is finally evaluated by the client.

Deliver, install and use the instrument!

Chapter 17
SENSOR AND ANTENNA
RESOLUTION IMPROVEMENT

Chapter 17
Sensor and Antenna Resolution Improvement

There are a number of different forms of spatial sensor that are used to either detect the presence of some other object, or perform imaging of objects. All of these sensors have some sort of response curve that reflects their sensitivity to a target at varying distances or angles off center. In *Figure 17-1* the sensor is defined as some sort of generic electro-optical sensor; e.g., a phototransistor or photo-op-amp. It could just as well be an ultrasonic imaging sensor, a radio antenna, or a radar set. When researching this article the electro-optical sensor was easily at hand, so was used.

The response of the sensor in *Figure 17-1* is highest immediately opposite the lens, at point X_o, as indicated by the peak voltage V_p. If the sensor and target translate relative to each other along the X axis, then the signal voltage (V) will rise from near zero to V_p, and then decrease to

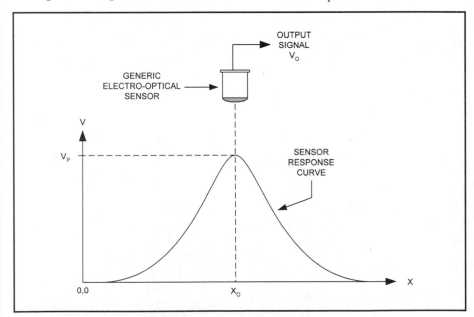

OUTPUT
SIGNAL
V_O

GENERIC
ELECTRO-OPTICAL
SENSOR

V

V_P

SENSOR
RESPONSE
CURVE

0,0

X_o

X

Figure 17-1.
A generic electro-optical sensor and its response curve.

near zero again as the target passes through the field of view. Although the curve in *Figure 17-1* looks suspiciously like a bell-shaped curve, actual curves might be shaped a bit differently, but the general form is correct.

Figure 17-2 shows a somewhat more practical situation found in many circuits. The response curve is the same, but there is a *sensitivity threshold* below which there is little or no output. This threshold might be generated by the brightness or size of the target, by ambient lighting, or be an intentionally set circuit value. In the latter case, it is common to see such thresholds to combat the effects of noise. In essence, the threshold level is a signal-to-noise ratio issue. The effect of the threshold is to improve the field of view, increasing the resolution, by narrowing the range of values of V that will be accepted. Even as improved by threshold detection, however, the field of view may be too great to provide adequate spacial resolution.

The *resolution* of the sensor is a measure of its ability to separate two equal targets. If the resolution is not matched to the objects being measured, then an ambiguity occurs. This is seen in popular films such as *Top Gun* where what the F-14 aircrew thought was two enemy fighters suddenly broke out to four dirty smelly bad guys, much to the dismay of the good guys. The radar resolution apparently wasn't able to distinguish two fighters flying close together.

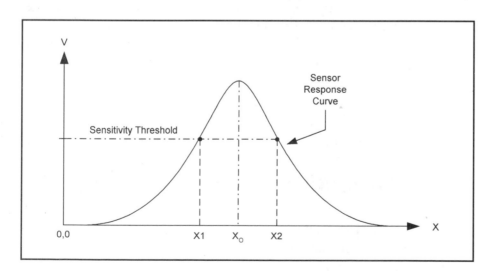

Figure 17-2. Sensor response curve. A practical threshold limit narrows the field of view, but not sufficiently for high-resolution operation.

There are a number of practical situations where sensor resolution can cause problems. For example, if a photosensor is used to count manufactured products coming down an assembly line, poor resolution means the items being counted would have to be further apart in order to avoid a miscount. In imaging systems the resolution can determine the smallest object that can be properly displayed. Poor resolution might cause a distortion of the object being imaged, or completely miss some important feature. It simply wouldn't do for a medical imaging system to miss your kidney stone! In robotics, if an electro-optical sensor is used as the eyes of the robot device, then poor resolution can hamper its ability to perceive and negotiate its environment. I recall one smart lad who built a robot that tooted around the room, and when its internal battery dropped below a certain point, then it would search the walls of the room for an electrical outlet. It did so by comparing a pattern of an outlet stored in memory with what it saw in the room. Poor resolution might cause it to mistake Aunt Annie's belt buckle for the outlet, and wouldn't that cause a family row!

Figures 17-3 and *17-4* show these effects in graphical form. In *Figure 17-3* a single target is in the field of view (FOV) of the optical sensor. Assume that the sensor moved left to right across the target, producing the output voltage shown. It doesn't matter whether the sensor or target moves, so long as there is relative motion between the two along the X-axis. Unfortunately, the sensor FOV, which determines the resolution, is too broad, so the target appears to be smeared in the X axis. The size and exact location data are thus distorted.

The situation in *Figure 17-4* shows two targets in a similar situation. Again we suppose that the sensor translated left to right across the two targets. Because both targets fall inside the FOV simultaneously, they will appear smeared, but (maybe) with a small dip to indicate the space. If the dip is too small to detect, then it will not be seen.

Sensor Resolution Improvement (SRI)

The sensor resolution cannot be improved without redesigning the device. In some cases, the laws of physics might prohibit further improve-

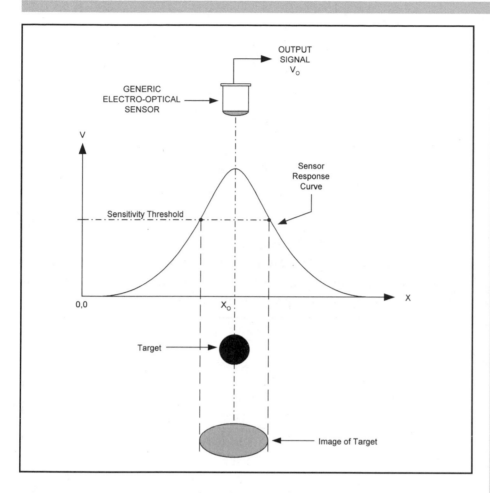

Figure 17-3.
On a single sensor
the fact that the target
falls within the field
of view smears the
image of the target.

ment. But there is something that can be done to correct the problem. This method is derived from a radar technique called *monopulse resolution improvement* (MRI). There are two versions of the circuit. One uses analog methods, but requires two sensors (a case of two being much better than one). This first approach can also be implemented in a computer version. A related method can be implemented using a digital computer, but proves difficult in analog circuitry.

In radar the target is illuminated with two adjacent coplanar antennas, and the returned signal processed in a special way. Assume two signals are V1 and V2. If we create sum (V1 + V2) and difference (V1 - V2) signals from these raw signal, then we can accomplish a tremendous resolution improvement. The equation is:

$$V_o = \frac{V1 + V2}{k + ABS(V1 - V2)} \qquad (17\text{-}1)$$

Where:

> V_o is the resolution-improved signal
>
> V1 and V2 are the input signals
>
> ABS indicates the absolute value of V1-V2
>
> k is a small full-scale constant

By dividing the sum by the absolute value of the difference at each point along the X-axis we create the resolution-improved signal V_o. The factor k is a small value constant set to prevent a divide-by-zero error when V1 = V2, or an extremely high value when V1 and V2 are very close in value. The value of k is set to produce full-scale output when V1-V2 = 0.

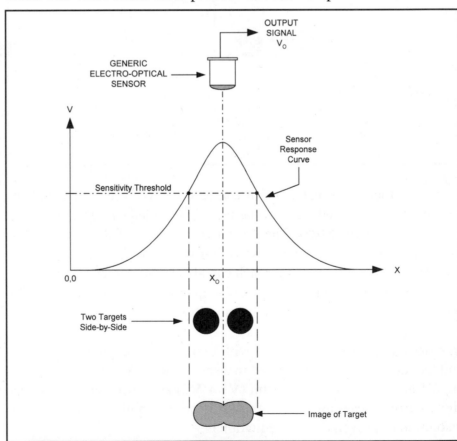

Figure 17-4. Resolution problems prevent breaking out the two targets, both of which fall within the sensor's field of view.

Initial Trials

The two-sensor approach was modeled first in an *Excel* spreadsheet. For the first attempt, a curve similar to those in *Figures 17-1* through *17-4* was converted to numbers and entered in successive cells of a single column. Each cell represented another increment along the X-axis, while the value in the cell represented the signal voltage V at that point. When the method was applied and graphed, there was a tremendous improvement in the resolution of the hypothetical sensor. The time came to "cut metal" and build a real circuit model.

Actual Circuit Trial

Figure 17-5A shows the actual test setup used to acquire data. Two electro-optical sensors were obtained. The optical sensors were Burr-Brown OPT-101 devices. These sensors are operational amplifiers with a photodiode device built-in to the transparent 8-pin DIP IC package. The two OPT-101 devices were spaced 10 mm apart so that their cones of acceptance overlapped (which is also the mini-

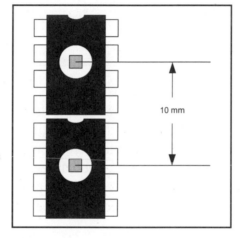

Figure 17-5A. Two-sensor test device made by mating a pair of Burr-Brown OPT-101 devices.

mum possible X-axis separation due to the size of the IC packages). This mounting was convenient because the two devices would fit nicely end-to-end in a single 16-pin DIP socket. The midpoint between the two OPT-101 devices corresponds to X_O, while the distance between them corresponds to DDX; any particular point along the path between S1 and S2 is designated X_i.

The target was a red light-emitting diode (LED) mounted on a movable stage on a microcrometer gear rack device that measured distance of travel in millimeters (*Figure 17-5B*). The initial position of the LED target was set so that it was outside the FOV of both S1 and S2. The LED was then advanced 1 mm at a time until it had traversed the entire dis-

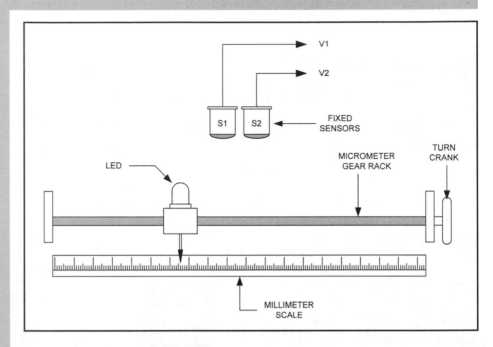

Figure 17-5B.
Test fixture allowed
the LED target to be
translated along the
X-axis in front the
sensors.

tance from the left-most extent of the FOV of S1 to the right-most extent of the FOV of S2. The output voltages (V1 and V2) from S1 and S2 were measured with a 3½-digit digital voltmeter (DVM) at each 1-mm interval. Those data were then entered into an Excel spreadsheet and plotted on a chart. The curves shown in *Figure 17-6* represent actual results from this experiment, rather than simulated results. The experimental and modeled results were the same.

All of the curves are shown in *Figure 17-6*—V1, V2, V1+V2, V1-V2, ABS(V1-V2) and the resolution—improve V_o. The V1 and V2 curves represent the normal FOV of the OPT-101 devices, and therefore the effective resolution of the devices. Note how much narrower V_o is, compared with either V1 or V2. This response curve would resolve much finer separations and produce superior images than either S1 or S2 alone.

Single-Sensor Method

The two-sensor method of SRI produces startling results, but at the cost of two sensors. It is easy to implement in analog circuitry, and can also be implemented in digital circuitry. Another approach uses a single sen-

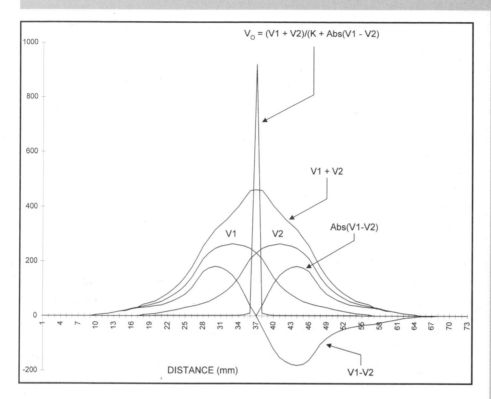

Figure 17-6.
Response curves of
LED/opto-amplifier
trial. V1 and V2 are
the outputs of
sensors S1 and S2.

sor and a look-ahead technique to synthesize V2. It is easily implemented digitally, but is somewhat difficult to implement in analog circuitry.

Assume that the values of V1 from the sensor are a series V_i in which each value represents the signal amplitude at sequential locations along the X-axis:

V1 V2 V3 V4 V5 V6 V7 V8 V9 V10 V11 ... V_{ith}

Each value of V_i represents a value of V1 in *Equation 17-1*. The corresponding value of V2 in the same equation is found by taking a subsequent value of V1 that is displaced a distance N (an integer) from V_i. Thus, in terms of *Equation 17-1*:

$$V1 = V_i \qquad\qquad\qquad\qquad (17\text{-}2)$$

$$V2 = V_{i+N} \qquad\qquad\qquad\qquad (17\text{-}3)$$

Equation 17-1 can be rewritten to the form:

$$V_O = \frac{V_i + V_{i+N}}{k + (V_i - V_{i+N})} \qquad (17-4)$$

When the V1 data from the experiment are plotted, the resultant curves are very similar to those of *Figure 17-6* and demonstrate very nearly the same degree of resolution improvement. It was also found that a limited amount of "tuning" of resolution can be done by selecting values of N. However, with the 1 mm spacing used in the experiment, values N > 5 showed essentially the same curves.

In both the single-sensor and two-sensor methods there exists the possibility of creating a selectable beamwidth sensor system. Signal V_O could be the narrow beamwidth signal, V1 + V2 can be the wide beamwidth signal, and V1 (or V2) can be the medium beamwidth signal.

Conclusion

The sensor resolution improvement (SRI) method is a variant on the monopulse resolution improvement (MRI) method used for many years in radar technology. It appears to have application in ultrasonic imaging (which resembles radar in basic approach), and any instrumentation problem where sensor resolution is an issue. It should also work nicely with a pair of identical antennas used in RadioScience Observing.

References

Brown, John M. Brown and Joseph J. Carr (1997). *Introduction to Biomedical Equipment Technology - 3rd Edition*. New York: Prentice-Hall

Carr, Joseph J. (1992). *The Art of Science*, San Diego: Hightext Publications, Inc.

Skolnik, Merrill (1980). *Introduction to Radar Systems*, New York: McGraw-Hill.

Stimson, George W. (1983). *Introduction to Airborne Radar Systems*, El Segundo, CA: Hughes Aircraft Co.

Chapter 18
LONG PERIOD VELOCITY TYPE SEISMOMETERS

Chapter 18
Long Period Velocity Type Seismometers

Allan Coleman
(An Invited Chapter)

This chapter describes how it is possible to lengthen the natural period of a velocity type seismometer using electronic feedback. This chapter assumes that the reader has made, or has the knowledge to make, a horizontal long period (5-15 seconds) velocity type seismometer. A good article which describes to how make such an instrument for amateur use was published in *Scientific American*, July 1979 edition, in the column titled "The Amateur Scientist" (see Chapter 13 of this book also). It detailed the fabrication of an instrument as well as how to record the ground motion it detects. The device can be adapted to that seismometer design.

Technical Description of the Prototype Seismometer

The design of the prototype horizontal seismometer is similar to the one written about in the *Scientific American* magazine's article referenced above. The major differences are as follows:

a) The seismic coil is larger, 4 1/2" OD x 1-1/8" wide. The coil is wrapped on a wooden former 2-3/8" diameter. The magnet wire is not all the same size, as different gauges were pieced together to make a longer length (total coil resistance = 3.7 kohms)—gauges ranged in size from #34 to #38, depending on what was found at the junkyard.

b) The seismic coil is free to move in a U-shaped magnet with an 1-1/4" gap, attached to the baseplate. If there is any magnetic material, (carpentry nails, a door hinge, rebar in the supporting cement slab, etc.) adjacent to the seismometer, the pendulum-mounted magnet will be attracted to it and not respond accurately to the ground motion.

c) The seismic coil is mounted at the end of the boom. Heavy-gauge wire carries the signal along the boom, steps down to a much smaller gauge wire to jump from the boom to the support column, then back to a heavy gauge again to the seismometer preamp. The weight of the coil did not necessitate any additional mass attached to the boom.

d) The feedback magnet is attached to the base. NOTE: If you are going to have the seismic signal-generating magnet mounted on the pendulum, then have the feedback magnet fitted to the pendulum also. That way the magnets are not fighting each other and upsetting the pendulum performance.

e) The feedback coil assembly, mounted on the pendulum, should be located closer to the seismic coil than to the pivot end. No closer than 8" to the seismic coil because we need to minimize any effect of the feedback signal on the seismic coil signal.

f) The pendulum length, from pivot point to seismic coil, is 21".

g) Preamp electronics are more complex because of the different function requirements.

Technical Description of The Period Lengthening Device

Mechanical

The feedback coil is made on a PVC plastic former as shown in *Figure 18-1*. The coil was made from 38 AWG enameled magnet wire, weighing approximately 2 oz., with a resistance of about 1500 ohms. All materials that hold the coil in place should be nonmagnetic, otherwise the magnet will suck up to it. The feedback coil is connected to the preamp

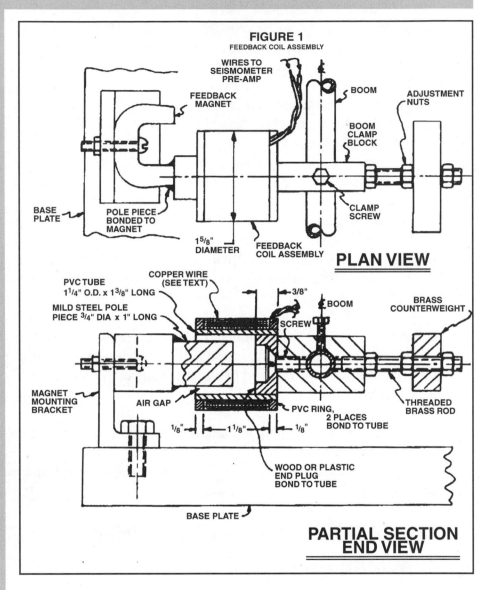

FIGURE 1
FEEDBACK COIL ASSEMBLY

WIRES TO
SEISMOMETER
PRE-AMP

BOOM

ADJUSTMENT
NUTS

FEEDBACK
MAGNET

BOOM
CLAMP
BLOCK

BASE
PLATE

POLE PIECE
BONDED TO
MAGNET

CLAMP
SCREW

$1^{5}/_{8}$"
DIAMETER

FEEDBACK
COIL ASSEMBLY

PLAN VIEW

PVC TUBE
$1^{1}/_{4}$" O.D. x $1^{3}/_{8}$" LONG

COPPER WIRE
(SEE TEXT)

3/8"

BRASS
COUNTERWEIGHT

MILD STEEL POLE
PIECE $^{3}/_{4}$" DIA x 1" LONG

BOOM

SCREW

MAGNET
MOUNTING
BRACKET

AIR GAP

PVC RING,
2 PLACES
BOND TO TUBE

THREADED
BRASS ROD

$^{1}/_{8}$"

$1^{1}/_{8}$"

$^{1}/_{8}$"

WOOD OR PLASTIC
END PLUG
BOND TO TUBE

BASE PLATE

**PARTIAL SECTION
END VIEW**

Figure 18-1.
Seismometer
mechanical detail.

with the same wiring scheme that's used for the seismic coil, with some flexible fine wire adjacent to the pendulum pivot point. For whichever item is attached to the pendulum, either feedback magnet or coil, the cantilevered mass MUST be counterbalanced to prevent a rotational torque on the boom. See the setup illustrated in *Figure 18-1*. It has a brass counterweight on the opposite side of the boom. Threaded rod allows for convenient adjustment.

The magnet (made by General Hardware Mfg. Co., Inc.), whose proportions can be taken from *Figure 18-1*, is rated with a holding force of 22 lb. and is made of Alnico. To the magnet was bonded a mild steel pole piece to help shape the magnetic field within the coil. An air gap exists between the sides of the pole piece and tube bore. A 3/8" space between the end of the pole piece and end plug allows for ample movement of the pendulum.

Electronic Circuit

The circuit may be broken down into two sections: the seismometer preamp (*Figure 18-2*) located adjacent to the seismometer; and the filter circuitry (*Figure 18-3*) located adjacent to the recorder. The seismic signal is fed into a difference amplifier (IC1) and then amplified 100X (IC2) before going to another signal amp (IC4). Depending on the type of recorder used it may be necessary to eliminate IC4 from the circuit and alter the resistor values as connected to pins 2 and 6 on IC2. To eliminate problems with high-frequency noise, the low-pass filter (IC5) is used and the high-pass filter (IC6) keeps signal drift in check.

Electronic feedback is achieved by monitoring the voltage generated by the seismic coil. The signal is cleaned up by a low-pass filter and sent to a differentiator (IC3A). The differentiator is used to produce a signal that is proportional to the rate of change of the input signal, so with a constant DC level the output of the differentiator is zero. As the rate of change increases, the output voltage also increases. The input signal is differentiated for up to 100 seconds—the RC time constant. The differentiator's output is converted to a current (IC3B) sufficiently large enough to power the feedback coil, thereby slowing down the pendulum motion.

Pendulum Responses of The Prototype Seismometer

Figure 18-4 shows different pendulum responses as they were made on a strip-chart recorder. NOTE: The signal amplitude is not supposed to be

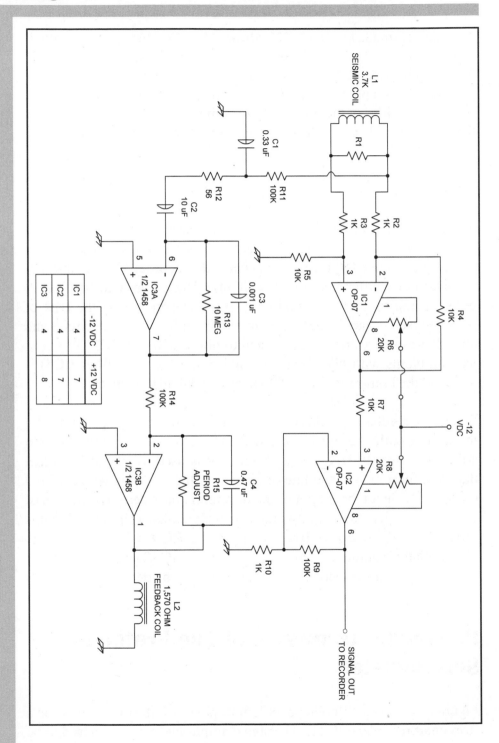

Figure 18-2.
Front-end circuitry.

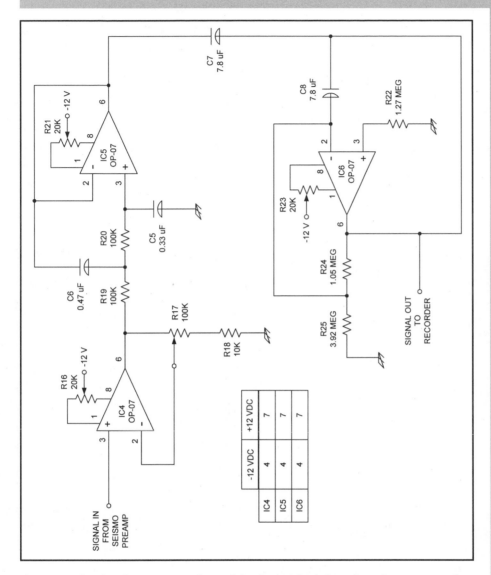

Figure 18-3.
Filter circuitry.

the same in the three examples—it's the period duration that we need to review. In *Figure 18-4A* the pendulum was free to swing with no damping at all. The natural period for the seismometer was originally adjusted to 10 seconds, but was shortened to 6.5 seconds to show the possible effects of period lengthening on a smaller, simpler-designed instrument.

A shunt resistor (R1) of 1 kohm was put across the seismic coil. The pendulum was tapped with a finger to observe how long it would take for it to come to a rest. *Figure 18-4B* shows that it took 18 seconds to return to its rest position with minimal overshoot. This test was repeated three times, as shown.

The feedback circuit was hooked up to the feedback coil and the results are presented in *Figure 18-4C*. The pendulum was clamped to a travel limit stop, about 1/2" off center, and released. The output signal seen at pin 6 on IC6 shows that the combination of damping resistor R1 and period adjustment resistor R15 presently selected gives the required damping with a 35-second period.

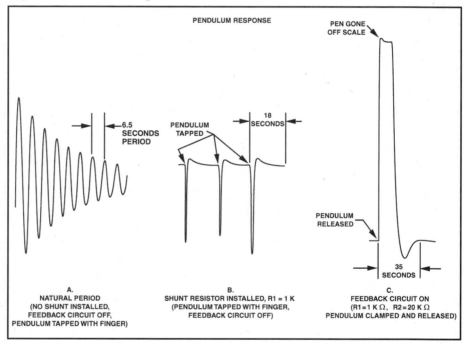

Figure 18-4.
Actual signals.

Electronic Period Adjustment

First build the circuits shown in *Figures 18-2* and *18-3*. Lay out the boards where it is easy to swap-out resistors R1 & R15. To achieve the desired results with the instrument that you apply this device to, some experimentation should be expected. For resistor R15, the value chosen for the prototype was 20 kohm. When this value increases, the period is lengthened because the feedback coil receives more current to hold back the pendulum. This is the equivalent of adding more mass to the pendulum. As the period lengthens (by increasing the resistance of R15) the shunt resistor R1 value needs to be decreased to compensate for the added inertia, thereby providing more damping. When R1 resistance is decreased, it robs the signal going to IC1 and IC3A.

When the electronics are powered up while connected to the coils and the pendulum starts to oscillate on its own, it indicates that the feedback coil is wired backwards. Simply reverse the polarity of the wires connected to the feedback coil.

Comments

If you want to try for extra long periods of 60 seconds or more it may be necessary to add a noninverting signal amplifier into the circuit between the pickoff point at R1 (point A) and the 100k resistor of the low-pass filter. This amp will help boost the voltage signal going to the differentiator because R1 will become a very low-value resistor, robbing signal strength from the seismic coil. The 10-megohm resistor on IC3A should be increased to 15 megohm to lengthen the differentiator time constant to 150 seconds. Either horizontal or vertical velocity seismometers can have the period lengthened with electronic feedback. Long period vertical seismometers can be a real nightmare to operate due to the fact that they are very sensitive to barometric pressure changes. I have been informed that the effect is proportional to the square of the period; that is to say, a 30-second vertical is 900 times more sensitive than a 1-second vertical. The pendulum "floats" in the air. Even building such an instrument out of extremely dense material will not help. The seismometer must be put in an airtight, rigid-wall enclosure. The long period horizontal seismometer is mostly affected by tilting. An enclosure for it may be best left vented to the atmosphere, otherwise the warping of the enclosure will cause the base to warp, tilting the instrument.

I hope that you achieve the same results as seen with the prototype seismometer. *Please let me know of any problems that you find with this design so the original document can be corrected.*

Chapter 19
Improving VLF Performance of Your Shortwave Receiver

A number of RadioScience Observing activities take place on the VLF and LF frequency bands. Some of the receivers used are specialty models, while others are homebrew. Still others were originally designed for other frequencies, but have VLF and/or LF in the bargain. Modern shortwave communications receivers are often advertised to operate well down into the Very Low Frequency (VLF) frequency range. Ads are seen with specifications of a lower end of 100 kHz (0.1 MHz), 50 kHz, 30 kHz or even 10 kHz. However, one thing that too many of those receivers have in common is that the performance of the receiver, which might be quite exciting on HF shortwave bands, is pukey on the VLF bands. Nothing ever seems to pop out of the noise level!

Other people like to "DX" the VLF bands for other purposes. Many people have built VLF receivers to track solar events. These radio astronomy amateurs look for the solar events that cause *Sudden Ionospheric Disturbances* (SIDs). The SID will usually cause a serious fadeout of the HF shortwave bands, lasting hours, but will also *increase* the strength of VLF stations during daylight.

Amateur radio astronomers monitor VLF frequencies between 20 and 40 kHz using "tuned radio frequency" receivers equipped with 1N60 voltage doubler half-wave rectifiers and a 220 mF filter capacitor.

Other people like to monitor standard time and frequency stations such as the 60 kHz WWVB signal in the USA. WWVB is a very high accuracy time and frequency station operated by the *National Institute of Standards and Technology* (NIST, formerly NBS) at Fort Collins, CO, co-located with the HF WWV stations. This signal can be used to operate very high accuracy frequency standards, or to make highly accurate clocks.

In this chapter you will find several approaches to improving the performance of VLF receivers. Also, be sure to look into the possibility of making a shielded loop antenna for VLF work. Details of suitable antennas can be found in my books *Receiving Antenna Handbook* (Universal Radio Research) and *RadioScience Observing, Volume I* (Howard W. Sams/PROMPT). For VLF reception, try using a loop with 64 to 140 turns of wire. Shielding is necessary to prevent stray local electrical fields from power lines and appliances from interfering with reception. Although 60 Hz is a long way from 60 kHz, there is enough energy in the power lines to have significant harmonics well above 100 kHz, and when a sensitive receiver is being used those harmonics are terribly loud.

Reducing AM BCB Interference on VLF

The AM broadcast band covers 540 to 1700 kHz, while the LF and VLF frequencies are those frequencies below 540 kHz. Therefore, one should not expect interference from local AM broadcast stations while noodling around in the "down unda" bands, right? No, not by a long shot. There are two basic reasons. First, the AM stations may tend to be *very* local. Second, radio stations tend to be high powered. One location where I lived was known locally as "Intermod Hill" because there were a large number of radio transmitters up there, including two FM BCBs (50 kW each) and one 5000-watt AM BCB station. Living just a few blocks from that 5000-watt gigablaster tore a hole in every receiver I owned. With such strong field strengths, almost all receivers overloaded.

Incidentally, the way I graded receivers in those days was how free they were from such interference. And it wasn't always the most expensive model that was the most free of problems.

A way to solve the problem is to use a low-pass filter (LPF) ahead of the receiver that has a cutoff frequency somewhere below 540 kHz. *Figure 19-1* shows a 7-element LPF with a 0.1 dB ripple factor, designed for

50-ohm input and output impedances. The LPF consists of four inductances (L1 - L4) and three capacitors (C1 - C3). Each capacitor is made up of two parallel capacitances in order to obtain the correct capacitance. Those capacitors should be a relatively stable, 5% tolerance type, such as the *Panasonic* B-Series polyester or V-Series metallized film capacitors. For receive-only applications, the voltage rating is not an issue, so 50 WVDC capacitors in those series will work nicely.

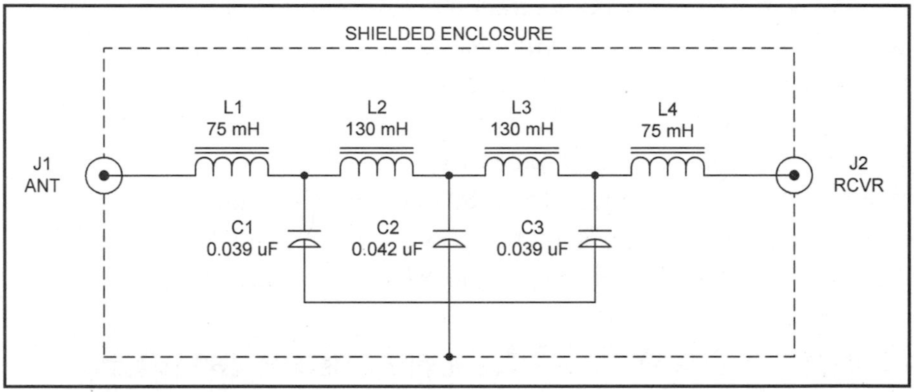

Figure 19-1. 120 KHz low-pass filter circuit. The same circuit can be used at any other design frequency (see text).

The inductance values shown in *Figure 19-1* are for the VLF version of the tuner, for which a 120 kHz -3 dB cutoff frequency is specified. The 75 mH inductors are *Toko* TK-4256, while the 130 mH inductors are *Toko* TK-4262. If you want to make the LPF work for a different cutoff frequency, then consult *Table 19-1* for component values:

Table 19.1. LPF values for different cutoff frequencies.

BAND	L1/L4	L2/L3	C1	C2	C3
<300 kHz	33 mH	56 mH	0.015 mF	0.0167 mF	0.015 mF
<500 kHz	18 mH	33 mH	0.009 mF	0.01 mF	0.009 mF

The layout of the LPF circuit should follow "good low-frequency RF practices," which means that you want at least a small separation between the inductors. Construction on perforated board is quite satisfactory. My initial version was built "rat's nest" style on printed-circuit perfboard. That form has printed-circuit connection points on a 0.1" x 0.1" pattern, so is easy to solder to and use. If you wish to make a printed-circuit board, a pattern in provided in *Figure 19-2*.

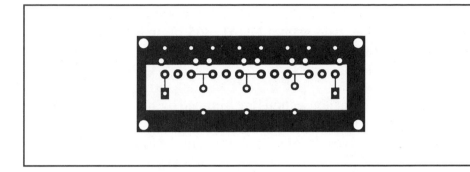

Figure 19-2.
Printed-circuit
board layout.

By the way, the same circuit (*Figure 19-2*) and PCB can be used for any frequency. This LPF is somewhat universal. To find the values of the components use the following formulas:

$$L1 = L4 = 9.4/F_{MHz} \qquad L2 = L3 = 16.68/F_{MHz}$$

$$C1 = C3 = 4538.9/F_{MHz} \qquad C2 = 5008.3/F_{MHz}$$

Where: F_{MHz} is the frequency in megahertz.

The coils can be made from T-37-xx toroidal cores, or selected from the Toko coils. Select fixed coils with a 5 mm (0.2-inch) pin spacing...which is nearly all of them in that catalog.

Preselector/Preamplifier Project

There are two general problems with the VLF band in even some very good HF "general coverage" shortwave receivers: *sensitivity* and *front-end selectivity*. In both cases, there is a tremendous lack of capability in many receivers; both are addressed in this project. *Figure 19-3* shows a three-band VLF preselector and preamplifier for use ahead of a receiver (i.e., between the antenna and the receiver antenna terminals).

The circuit of *Figure 19-3* has two major sections: preselector tuning (*Figure 19-3A*) and amplification (*Figure 19-3B*). The preselector tuning consists of a dual-section, reactance-coupled parallel L-C tuning network. The main tuning capacitor (C1A, C1B) is a two-section ("dual") AM broadcast variable. Most of those capacitors are 365 pF per section (be careful—some models have an oscillator section that is less than 365 pF per section...you want a DUAL 365 pF unit). Also useful are dual 380 pF, 400 pF, 440 pF or 500 pF capacitors, all of which I've seen in recent catalogs.

The main tuning capacitor sections are trimmed by small variable capacitors (C2 and C3). For this particular project I selected the *Sprague-Goodman* 10-mm type. These capacitors are designed for top adjustment, so are good for cases where the printed-circuit board is mounted inside a cabinet (the recommended approach).

The three inductors in each side of the tuning network (L1/L2/L3 and L4/L5/L6) are switch selected. The switch is a 2P3T rotary switch; i.e., it has two poles and three positions. These are a little hard to find in some stores, but the 3P3Ts are much easier, and will work well (just ignore the third set of contacts). The inductors are arranged so that in position "A" all three are connected in series; in position "B" two are connected in series; and in position "C" only one coil on each side (L3 and L6) are connected into the circuit. The values for these coils are:

COIL	VALUE	NUMBER
L1/L4	56 mH	TK-1721
L2/L5	56 mH	TK-1721
L3/L6	33 mH	TK-1724

The two halves of the L-C tuned circuit are coupled by a small mutual reactance. Although either capacitive or inductive reactance can be used for this purpose (although in different ways), the method used in *Figure 19-3* is the capacitive version. Capacitor C4 is used to couple energy from one side of the L-C tank circuit to the other. A 10 pF disk ceramic or silvered mica unit will suffice. The idea is to get as small a value that will couple the two sides without allowing too much interaction. Some people prefer to use a small (<12 pF) variable capacitor in place of C4.

The tuning ranges with the coils shown were measured for the various positions of S1 as follows:

A	20.8 - 36 kHz
B	25.2 - 44.4 kHz
C	39.1 - 70.8 kHz

You can alter the frequency range by selecting different coils from the same series in the *Digi-Key* catalog, or from some other source (including winding your own, if you are so inclined). Keep in mind that the frequencies are approximate and will differ from the frequencies found from the equation (on your calculator) because of two reasons: first, the inductances have a tolerance associated with them; and second, the inductors have significant capacitance. The values will be within 15 percent or less, but if the frequency that you desire is close to the margin for some particular coil, then be prepared to use a different coil. If you chose to build the circuit for a different set of frequencies, then follow this rule of thumb: *the frequency change is proportional to the square root of the inductance change*.

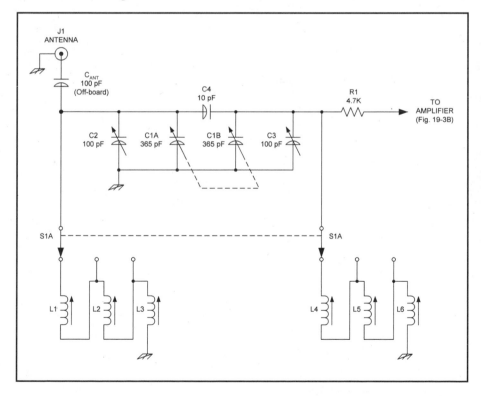

Figure 19-3A.
VLF preselector
tuning circuit.

The signal from the tuned circuit is coupled into a two-stage amplifier (*Figure 19-3B*) consisting of three transistors (Q1, Q2 and Q3). Transistors Q1 and Q2 are connected into the Darlington amplifier configuration, so offers a very high input impedance. This stage produces some power gain, but little voltage gain. If it were a common-emitter circuit, rather than a common-collector circuit, the gain would be huge (i.e., the product of the betas of Q1 and Q2). Because Q1/Q2 is an emitter-

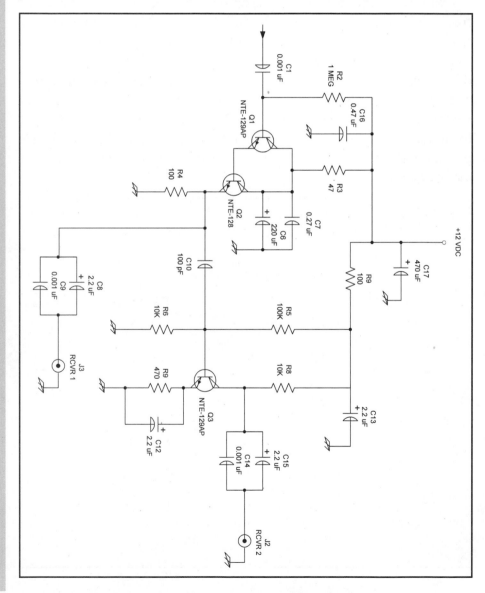

Figure 19-3B.
VLF preamplifier
circuit.

follower (common-collector) amplifier, signal is taken from the emitter of Q2 and fed to two destinations: a low-gain output (OUT1) and the input of the final amplifier stage (Q1).

The circuit of Q1 is a common-emitter amplifier. It uses the standard voltage divider form of biasing, with an emitter resistor to stabilize the circuit. The output of this amplifier is taken from the collector, and fed to the high-gain output (OUT2).

The transistors selected for this project are the NTE-128 and NTE-129AP (also, the ECG-128 and ECG-129AP will work as well). These parts are available from many parts distributors. Neither transistor is particularly critical, however. Q1 and Q2 were replaced with 2N4401 in one version, and worked the same as the selected transistors. Transistor Q2 can be any "audio output/video driver" NPN silicon device with a gain-bandwidth product of 100 MHz, and a beta gain around 90.

The power distribution network takes the +12 VDC power supply and sends it to the two stages. In both stages, there is a small-value resistor (R3 and R9) used to improve decoupling between stages. The actual decoupling is performed by capacitor pairs C6/C7, C11/C13 and C16/C17. In each case, a pair of capacitors, one low-value and one a high-value electrolytic, is used. The high-value capacitors are used to smooth power supply variations and reduce any power supply noise, as well as decoupling low-frequency feedback through the DC power supply. The low-value capacitors are used for high-frequency decoupling.

A printed-circuit board layout for this project is shown in *Figure 19-4*. This board is a simple single-sided design, so is well within the capabilities of homebrew PCB kits.

Alignment of the circuit is relatively straightforward, although it requires a signal generator. The signal generator could be an RF generator that goes down to 10 kHz (a few exist), or an audio generator, or function generator that has a low-impedance sinewave output (50 ohms or 600 ohms)...which is about all of them.

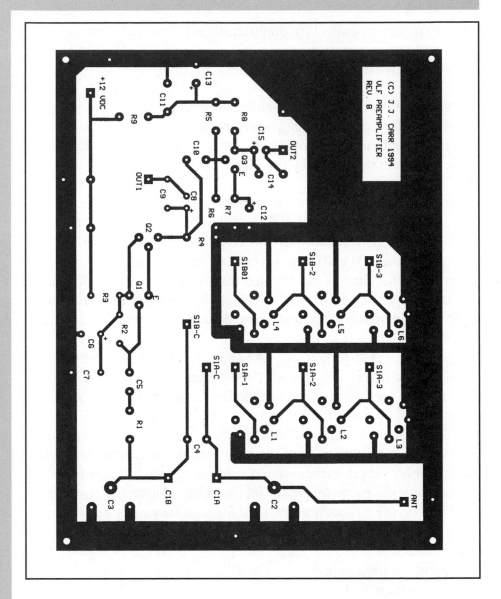

Figure 19-4.
Printed-circuit board
layout

Alignment begins with the tuning capacitor (C1) fully meshed, the bandswitch set to position "D," and either an AC voltmeter or oscilloscope monitoring the amplified output (OUT2). Alternatively, connect the preselector to the receiver and find the minimum frequency point on its dial. Adjust the signal generator frequency until a peak is found. In all likelihood, the first cut will show a "double humped" tuning characteristic because the two halves of the L-C network are not adjusted to the same frequency. Adjust L3 and L6 for a peak response. Next, adjust C2 and C3 for a peak response. These adjustments are interactive, so do it two or three additional times to make sure that the actual peaks are reached.

Next, set the tuning capacitor fully open (all the way unmeshed). Find the peak frequency by adjusting the signal generator. For this adjustment, leave the inductors alone and just peak C2 and C3.

Next, align band "B" in the same manner, except that the capacitors are not trimmed. That means you adjust L2/L5 only. Finally, adjust band "A" by following the same procedure for L1/L4.

Variations on the Theme

There are two variations on the circuit that may interest some readers. First is varactor tuning for those who can't find the dual tuning capacitors, or don't want to use the capacitor. In this circuit (*Figure 19-5A*), an AM BCB varactor (voltage-variable capacitance) diode is used to replace the capacitor.

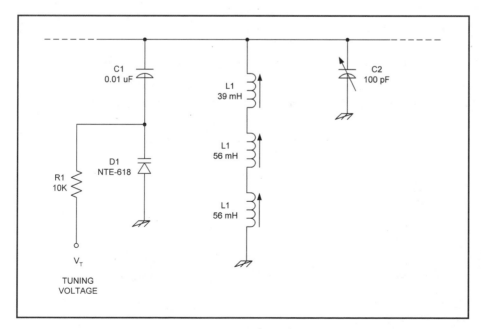

Figure 19-5A.
Varactor tuning.

The varactor selected is the NTE-618, which has a capacitance of a few picofarads up to 440 pF as the applied reverse bias tuning voltage V_T is changed from 12 VDC down to 1.00 VDC. Capacitor C1 is used to provide DC blocking that keeps the tuning voltage (V_T) from affecting (or being affected by) the inductors and following circuitry.

The tuning characteristic for the VLF preamplifier, found during a bench test, is shown graphed in *Figure 19-5B*. With this combination, the tuning ranged from 20.5 to 28.5 kHz when the trimmer was set to 45 pF and the inductances were set to 116.5 mH.

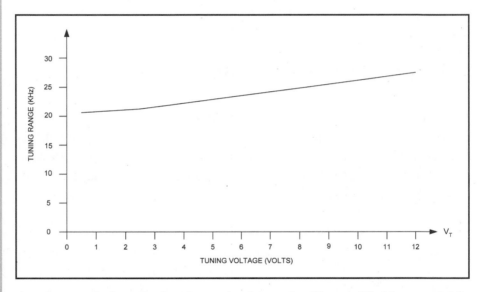

Figure 19-5B. Sample tuning characteristic.

Another variation on the theme is shown in *Figure 19-6A*: a variable attenuator circuit. This network is inserted into the circuit by replacing capacitor C4 in *Figure 19-3A* with this network (although it will have to be built off-board). The attenuation occurs because of the action of the hot carrier Schottky ("PIN") diode, D1. I selected an NTE-553 diode for this job because of easy availability. A 3 volt peak-to-peak, 25 kHz signal was used to test the circuit.

The results of the test are shown in *Figure 19-6B*. Notice that nearly all of the drop occurs between 0 and 1 volt DC. Although it is a straight line, I suspect that the shape would be curved if more data points were collected.

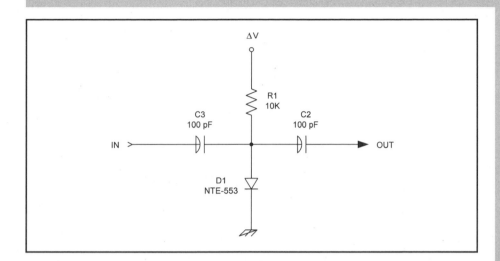

Figure 19-6A.
PIN diode
attenuator.

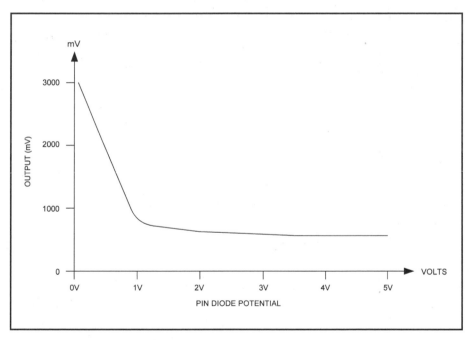

Figure 19-6B.
Attenuation
characteristic.

Conclusion

The VLF bands of many otherwise fine shortwave communications "general coverage" receivers are often woefully inadequate. The ideas in this chapter will make some of those pukey performers work a whole lot better.

Chapter 20
ANTENNA TRANSMISSION LINES AND VSWR LOSS

Chapter 20
Antenna Transmission Lines and VSWR Loss

If you deal with VHF, UHF or microwave antennas and receiving or transmitting systems, then it is critical that you understand not only the role of coaxial cable loss, but also how VSWR measurements made at the receiver or transmitter end (as opposed to the antenna end) can be highly erroneous. This might lead you to a false conclusion regarding the quality of the system.

The Problem

All electrical sources have an internal resistance or impedance: batteries, the AC power mains, radio transmitters, and signal generators all have some value of internal resistance (*Figure 20-1*). In transmitters and signal generators that value is usually stated as "output impedance." The impedance could be any value $Z_S = R_S \pm jX$, where R_S is the resistive component and X is either a capacitive or inductive reactance. In RF circuits other than television the standard system impedance is 50 + j0, or simply 50 ohms resistive (not reactive component).

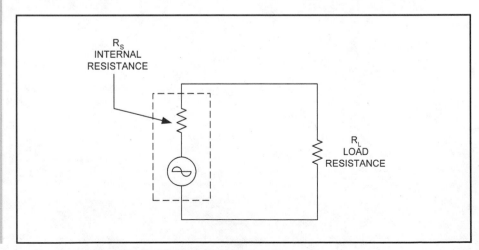

Figure 20-1. Source and load impedances must be matched for maximum power transfer.

A load will also have some value of impedance: $Z_L = R_L \pm jX$. Although transmitters and signal generators will control the output impedance so that it is purely resistive, and matches the accepted value of standard impedance (50 ohms, except for TV which is usually 75 ohms), the load may vary over quite a range of values of R and X.

It is rare to find a load connected directly to the output of a transmitter or signal generator. Most commonly, one expects to see a transmission line (*Figure 20-2*) between the source and load. The transmission line also has a characteristic impedance, which is denoted by Z_O.

One of the fundamental facts about connecting source to load is that maximum power transfer occurs when the load and source are matched; i.e., when $Z_S = Z_L$. If this is not the case, then not all of the power is delivered to the load. In a transmitter-antenna system, mismatch means not all of the available power is radiated as a radio signal. In a signal generator test setup, it means a possibly erroneous measurement (and, rarely, damage to the measuring equipment). The problem becomes more complicated in real situations because of the increased number of possible mismatches. It is necessary to match Z_S to Z_O, and Z_O to Z_L.

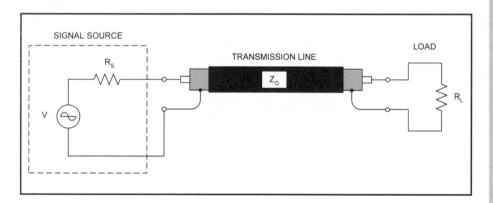

Figure 20-2. Transmission line connecting transmitter and load.

If a load is not matched to a source, then some of the RF power supplied by the source will not be absorbed in the load. It will be reflected back towards the source. Thus, we must contend with both the applied, or forward, power (P_F) supplied by the source, and the reflected power (P_R) rejected by the load.

Standing Waves

When the forward and reflected waves of an RF signal combine in the transmission line, they algebraically add, and set up a pattern of standing waves. In the case where $Z_O = Z_L$, there are no standing waves, and such a line is said to be "flat." But if $Z_O \neq Z_L$, then the standing waves emerge, producing a nonzero reflected power, and voltage nodes (and current nodes, incidentally) along the line. The voltage will vary from a maximum, V_{MAX} to a minimum, V_{MIN}. The maxima and minima are a quarter wavelength apart, and repeated maxima and repeated minima are half a wavelength apart. One implication of this situation is that an impedance connected to the end of a transmission line is repeated every half wavelength.

The reflected wave can be defined in terms of a *reflection coefficient* (Γ):

$$\Gamma \;=\; \frac{Z_L - Z_O}{Z_L + Z_O} \;=\; \frac{V_R}{V_F} \;=\; \sqrt{\frac{P_R}{P_F}} \qquad (20\text{-}1)$$

The standing waves are defined in terms of the *standing wave ratio*, which can be calculated from the reflection coefficient:

$$SWR \;=\; \frac{1 + ABS(\Gamma)}{1 - ABS(\Gamma)} \qquad (20\text{-}2)$$

The SWR can also be defined in terms of impedances:

$$SWR = Z_L/Z_O \text{ if } Z_L > Z_O \qquad (20\text{-}3)$$

or,

$$SWR = Z_O/Z_L \text{ if } Z_O > Z_L \qquad (20\text{-}4)$$

Finally, SWR can be defined in terms of power:

$$SWR \;=\; \frac{1 + \sqrt{P_R/P_F}}{1 - \sqrt{P_R/P_F}} \qquad (20\text{-}5)$$

Note: Although SWR can be measured using any of these properties, it is common practice to use the term *voltage standing wave ratio*, or VSWR, as synonymous with SWR. That usage will be observed herein.

Scenario

A technician was sent to check out a newly installed low-band VHF (30-50 MHz) communications antenna. He returned and told his boss that everything was in order because the VSWR was about 1.66:1, which was less than 2:1, so it met specifications. The Wisened Old Boss ("WOB", if you must have an acronym) was skeptical. He agreed that a VSWR of <2:1 was within the specification. But ol' WOB knew that low VSWR sometimes conceals deeper problems.

The antenna was mounted on a tower, and the tower was up on the crest of a hill. Altogether, there was 250 feet (76.2 meters) of RG-8/U 52-ohm transmission line. On further questioning, WOB discovered that the particular model antenna used on the system had a feedpoint impedance of 300 ohms resistive. That would infer a VSWR of 300/52 = 5.77:1, not 1.66:1 that the technician measured.

So what happened? The VSWR should be almost 6:1, but it only read 2:1. Why? The solution is in the loss of the transmission line. There are two basic forms of loss in coaxial cable: *copper loss* and *dielectric loss*. The copper losses result from the fact that the copper used to make the inner conductor and shield has resistance. The picture is further compounded by the fact that RF suffers skin effect, so the apparent RF resistance is higher for any given wire than the DC resistance. Further, the skin effect gets worse with frequency, so higher frequencies see more loss than lower frequencies.

The copper loss is due to the I^2R when current flows in the conductors. When there is a high VSWR at the load end, part of the power is reflected back down the line towards the transmitter. The effect of the reflected RF current is to increase the average current in the conductor. Thus, I^2R loss increases when VSWR increases.

Similarly with dielectric loss. This loss occurs because voltage fields of the RF signal cause problems. The simplistic explanation is that the voltage fields tend to distort electron orbits, and when those orbits return to their normal state some energy is lost. These losses are related to E^2/R.

As with copper loss, dielectric loss is frequency sensitive. The *loss factor* of coaxial cable increases with frequency. Let's look at two examples. *Table 1* shows a popular, quality brand of two 52-ohm coaxial cables: RG-8/U and RG-58/U. Note that the loss varies from 1.8 dB/100' (0.059 dB/meter) at 100 MHz to 7.10 dB/100' (0.233 dB/meter) at 1000 MHz. The RG-8/U cable is larger than the RG-58/U, and has less loss. Take a look at the same endpoints for the smaller cable: 4.9 dB/100' (0.161 dB/meter) at 100 MHz and 21.50 dB/100' (0.705 dB/meter) at 1000 MHz.

Clearly, the selection of the cable type is significant. At the FM BCB (100 MHz), the smaller cable would show 12.25 dB of loss. If a 100-watt signal is applied at the transmitter end, the ratio of loss is $10^{(12.25/10)} = 16.79$, so only about 6 watts is available to the antenna. The problem is even worse if the antenna is used for 900-MHz cellular telephones. In that band, the RG-58/U cable loss is 20 dB/100' (0.656 dB/m), so overall loss is a whopping 50 dB. The power ratio is $10^{(50/10)} = 10^5:1$, so only about 10 mmW makes it to the antenna.

If RG-8/U cable is used instead of RG-58/U, then the losses would be 4.5 dB at 100 MHz and 16.75 dB at 900 MHz. At the cellular frequencies in the 900-MHz band the loss factor will still be high—about 47:1—so only 2 watts would make it through the loss. The rest is used to heat up the coaxial cable. Fortunately, certain specialty cables are available with losses around 2.5 dB/100' (0.082 dB/meter) at 900 MHz. Such cable would produce about 6.25 dB of overall loss, or a ratio of 4.2:1. That cable would deliver nearly 24 watts of the original 100 watts.

The problem is also seen on receiver systems. Suppose that a 900-MHz receiver is at the end of a 250-foot transmission line. Further suppose that the signal is a respectable 1,000 μV, which in a 50-ohm load is -47 dBm. A loss of 6.25 dB would make the power level at the antenna terminals of the receiver -47 dBm - 6.25 dB = -53.25 dBm, or about 485 μV (still a reasonable signal). But if less coaxial cable is used (RG-58/U instead of the specialty grade), then the loss is 50 dB, and the signal at the receiver would be -47 dBm - 50 dB = -97 dBm. This level is getting close to the sensitivity limits of some receivers, or about 3.2 µV.

The example given above was from telecommunications, but can apply equally whenever RF is sent over coaxial cable. The cable TV, local area network, and other users of strictly landline RF also see the same loss effect. The correction is shown in *Figure 20-3*. There is a low-noise amplifier (LNA) placed at the head end of the transmission line. It boosts the signal before it suffers loss.

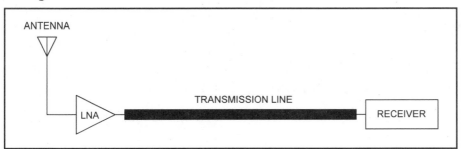

Figure 20-3. Receiver/ transmission line connection.

On first blush, it might seem easier to put the amplifier at the receiver end. But that doesn't work out so well because of two factors. First, there is an inherent noise factor in any amplifier. If the signal is attenuated before it is applied to the amplifier, then the ratio of the signal to the internal noise of the amplifier is a lot lower than if the signal had been applied before attenuation. So, while the signal would still be at the same level regardless of where the amplifier is placed, the all-important signal-to-noise ratio is deteriorated if the amplifier is at the receiver end. The second reason is that any lossy device, including coaxial cable, produces a noise level of its own:

$$F_{N(COAX)} = 1 + \frac{(L-1)T}{290} \qquad (20\text{-}6)$$

Where:

> $F_{N(COAX)}$ is the noise factor of the coax
> L is the loss express as a linear quantity
> T is the physical temperature of the cable in Kelvins

The linear noise factor due to loss can be converted to noise figure, which can be added to the system noise decibel-for-decibel.

Now let's return to the problem of the technician and WOB. The tables from the coaxial cable maker told WOB that the loss at 40 MHz is 1.2 dB/100', so the overall loss is 3 dB (halving the power). This loss is called the *matched line loss* (L_M). But, we also have to consider the Total Line Loss (TLL), which is:

$$TLL = 10 \, LOG\left[\frac{B^2 - C^2}{B(1-C^2)}\right] \tag{20-7}$$

Where:

B = Antilog L_M

$C = (SWR_{LOAD} - 1)/(SWR_{LOAD} + 1)$

SWR_{LOAD} is the VSWR at the load end of the line

WOB knew that the VSWR measured at the load end might be considerably higher than that measured at the transmitter end of the line. Given that L_M = 3 dB, B = LOG^{-1}(3) = 1.995. If $VSWR_{LOAD}$ = 5.77 (as is the case), then $C = (5.77-1)/(5.77+1) = 4.77/6.77 = 0.705$. Thus, the TLL is:

$$TLL = 10 \, LOG\left[\frac{1.995^2 - 0.705^2}{1.995(1 - 0.705^2)}\right] = 10 \, LOG\left[\frac{3.98 - 0.497}{1.995(0.497)}\right]$$

$$TLL = 10 \, LOG\left[\frac{3.483}{0.992}\right] = 10 \, LOG(3.51) = (10)(0.545) = 5.45 \, dB$$

The VSWR at the input end of the line, down the hill by the transmitter, then is:

$$VSWR_{IN} = \frac{B+C}{B-C} = \frac{1.995 + 0.497}{1.995 - 0.497} = \frac{2.492}{1.498} = 1.66:1$$

When he had made the calculations, WOB turned to the younger technician and told him that he had just improved his experience level—the
laddy was on his way to the wisdom that real experience provides. As
WOB headed out to the local honky-tonk to have his pint and a bowl of
grits, he told the boy, "Now you shag your lazy butt back up that hill and
impedance match that damn antenna!"

Appendix A
RADIO LUXEMBOURG EFFECT

Appendix A
Radio Luxembourg Effect

In *RadioScience Observing, Volume 1* I mentioned the rumored "Radio Luxembourg Effect" in which modulation transfer occurs between two signals in the ionosphere. I asked for information about this apparently poorly documented phenomenon. I received the letter below via e-mail from a reader, and pass it along for those who want to research the issue further.

"I congratulate you on the excellent book mentioned in...

With reference to Page 72 where the Luxembourg effect is being discussed, I should like to make some relevant comments:

I have studied Luxembourg effect since the 1960s but I have never seen any statements that European stations carrying the cross-modulated signals were being received in North America.

We have probably produced the only existing complete list of European LF/MF BC and commercial stations carrying on their carriers ICM produced modulation effects, the ICM products.

The total number of observed cases, i.e., European LF/MF stations carrying identified ICM products riding on their carriers, amounts to 161 stations, 78 of them affected by Warszawa 225/227 kHz, 11 affected by DDR 182 kHz, 12 by Loran C, 11 by HF Woodpecker and 39 by other ICM causing stations, mostly Russians but also other high-power ones. EBU also has made a series of ICM related measurements in the 1950s and 1960s, mostly of R. Luxembourg itself, but also with some other high-power LF BC stations.

[We have a report titled] "Observed Interference to LF and MF stations due to Ionospferic Cross-Modulation", 3rd March 1981

Also, EISCAT people in Sodankyla have been receiving a number of BC station audio signals on RF frequencies of 1 to 5 kHz!! In this case the ionosphere is somehow converting high-power BC station modulation to RF carrier signals. I have a couple of cassettes at home produced by Dr. Turunen of Sodankyla, where 1000 Hz time signal pips have been received on a radio frequency of 1 kHz. A lot of other material exists..."

V.K.Lehtoranta, OH2LX, Jokela, Finland

Appendix B
SCIENCE AND JUNK SCIENCE

Appendix B
Science and Junk Science

Copyright 1999 Howard W. Sams & Company
All Rights Reserved
Excerpted from *RadioScience Observing, Volume 2*
Prompt Publishing, Indianapolis, Indiana

"A Wise Man Proportions His Belief to the Evidence"

—Hume's Maxim

When wielded correctly, Ocham's razor can slice baloney exceedingly thin...so thin as to be transparent.

—J. Carr

Is Scientific Knowledge Superior to Other Forms?

Much of the material in this book, *RadioScience Observing, Volume 2*, is about science. No scientific training is necessary to read that material, although a high degree of interest is assumed. The purpose of this appendix is to make the reader aware of the processes behind the scientific method.

Scientific knowledge is superior to other forms of knowledge. This claim does not mean that knowledge of chemistry and physics is superior to knowledge of history or social issues. But rather, *knowledge that is obtained through a systematic, disciplined process—which we call the scientific method—is far superior to "knowledge" obtained through random guesses, hunches, intuition, or any form of non-systematic approach.* Much of what passes for "knowledge" in everyday life, is not knowledge in a scientific sense. Even when empirically derived ("experience"), unorganized observations cannot rise to the level of scientific knowledge.

One view of science is as a filter that helps us sort things out. Our knowledge of the real nature of things is often distorted by any number of factors resulting in a lot of "noise" on the "signal" of knowledge. The purpose of scientific method is to provide a filter that removes as much of the noise as possible, leaving only the underlying knowledge. And while the process isn't perfect, it is far above the random ranting of those who use other methods or no method at all.

Science provides a framework for, and examples of, "...effectively handling problems" [Crooke 1961] in a systematic way, so the scientific method should be the mainstay in the armamentarium of all thinking people. In 1942 Robert K. Merton proposed some "norms" that are characteristic of science (Casti 1989):

Originality. A hallmark of scientific studies is that they are original. With the possible exception of studies that are intended to replicate the work of an earlier study, the purpose of the scientific study is to somehow add something to the knowledge base. Studies that contribute neither something new, nor confirmation or refutation by replication of previous studies, are not true science.

Detachment. The only legitimate motive for scientific activity is the advancement of knowledge. Casti (1989) goes so far as to question whether or not scientific research conducted for industrial or defense purposes qualifies as "detached" and thus "science." The argument is that when monetary or other motives (e.g., ideology) impinge the scientific process, it deteriorates the goodness of the science involved. It is in this area that some of the most sensational scandals erupted. We disagree, and note that such factors are a caution, to be sure, but do not necessarily degrade scientific objectivity of the researcher.

Universality. The claims of scientific studies should be verifiable by anyone, anywhere, provided (of course) that they are equipped with the right apparatus and follow the same procedure. No special sources of information are permitted. Claims are based solely on the intrinsic merits of the data. No extrinsic factors such as religious, social, ethnic, racial or other prejudicial beliefs are permitted a place in the deliberation.

Skepticism. The scientific study proceeds from, and should be judged, on the basis of evidence alone. Nothing is accepted on faith; no one is trusted who has no data to support the claims being asserted.

Public Accountability. Central to scientific goodness, the basic quality of scientific studies is the matter of public accountability. The scientist places his or her data in the public square and lets the critical horde attack it for all they are worth. It is the accountability of peer review that makes scientific studies less likely to be poorly done than "private" studies.

A frequent cause of scientific failure comes from uncritical acceptance of a framework of ideology that causes leveling or sharpening of the data in order to support a preexisting result. Whether incidentally or by design, this problem is very widespread today (Sowell 1995).

Is Salt Water Taffy Being Distributed?

Junk science and revisionist history advocates usually want you to uncritically accept their point of view. Otherwise, you cannot be persuaded to their way of thinking. The tendency—and temptation—is for advocates to overinterpret favorable data, and either underinterpret or suppress unfavorable data. How do you separate trash from treasure?

How do you detect whether the offered food is nutritious, or merely salt water taffy? Several tactics help shed light on the truth.

First, *ask who is supplying the information*. There are advocacy groups that put out information to support or refute a position. Whether you deal with environmental issues, health food, alternative medicine, or the claims made for novel and bizarre radio antennas, you must ferret out the truth. One of the issues at hand is who supplied the information. Or, more precisely, how do they benefit if the information is accepted as true?

Newspapers and magazines often misinterpret scientific studies. While some of them are guilty of blatant advocacy, they are also afflicted by three pressures that force distortion: deadlines, the need to "scoop" the competition, and the fact that startling results "sell" better than the truth. Add to that the fact that journalists are rarely competent to interpret scholarly work, and the result is bad information being presented to the public. No one should accept popular media articles on any subject without independent verification.

Verification doesn't mean seeing the same material in two or more different media sources. Journalists often pick up stories from other publications. Seeing the same material in three publications could mean that one reporter wrote it for his or her newspaper (and got it wrong), and the next day (or next week) two more cribbed it for their papers.

Advocates attempt to gain credibility for their position by citing studies that support it. And if the study is a "scientific" or "university" study, then all the better. They will bolster their point by citing, by name, the author of the study...especially if the person quoted is famous, a well-known expert on the subject, or has the requisite credentials. "Doctor M. Weldon Sclotz of the University of Frabbitzville stated recently...."

Some people cite a well-credentialed expert who is found to not be expert in the matter under consideration. Having a doctoral degree, or winning a Nobel Prize, or being famous does not add credibility outside one's narrow field of expertise. Tom Clancy may be a real good military yarn spinner, but that does not qualify him as a military affairs expert.

When a study is cited by an ardent advocate, it is a good idea to look further. I often follow up on footnotes in order to judge the quality of research. In one case, an advocate's paper had plenty of footnotes, so on the surface looked like pretty good science. Footnotes are a paper's *bona fides*, they tell readers that the author did her homework. In that case, I found that a critically important footnote didn't even exist! The journal cited was real, but when the actual copy of the cited issue was located in a library, the article claimed in the footnote did not exist. Checking the annual index for that year, and the two years either side of it, revealed no articles or letters by the author cited in the paper. Mistake? Possibly. Deceit? I'd bet on it!

Critical footnotes need to be examined carefully to see:

      a) does the cited source actually exist

      b) is the source accurately quoted, and

      c) is the source *fairly used*?

The last criterion—fairly used—is especially important. Very often one finds that cited sources are taken out of context, or the conclusions of the study are twisted beyond recognition. If the author of the study would be surprised at the conclusions drawn, then it's a good bet that the study is not fairly used. While it is possible that an advocate could fairly draw different conclusions from those of a study author, that type of situation is always suspect until verified by the evidence. Junk science advocates often hitch their wagons to legitimate stars, but nearly everyone else finds their interpretation novel and bizarre.

A study can sometimes be verified by contacting its author. Ask the author if the published study is the latest information on the subject. Published academic studies tend to lag the state of knowledge. Some scholarly journals print the acceptance date of the paper, and that date is often a year or two earlier than the issue of the journal in which it appears.

Journals also frequently publish the study's funding information so that you can judge whether or not there may be any particular bias in the results. A typical citation might read "Funded under NIH grant XXX-XYZ-1230." If the funding organization is also an advocacy group or commercial entity, then be especially skeptical of the reported results (regardless of how well the group fits your own biases).

When you call the author of the study, ask about the methodology used. A lot of studies are terribly sensitive to the methodology used. Responsible researchers recognize that fact, and take it into account when reading the study, but journalists often don't. By understanding the methodology, it is possible to detect any biases, flaws, or limitations to the study.

Also find out whether the journal in which a published study appears is peer reviewed or merely editor reviewed. A peer review publication sends out every manuscript to one or more disinterested reviewers. They ask hard questions, provide critique and serve as general quality control. They point out weaknesses, and suggest areas for improvement.

Another factor often left out of popular expositions of scholarly research is critical qualifying factors. Studies are often tightly controlled as to the conditions being examined, and may not be valid in situations that depart from the controlled situation. It is often the case that studies will look at very narrow populations, and the results are not applicable to the population at large.

It may also be true that the interpretation of a study is different because critical information is left out of the report. Berkman (1993) tells of a report that the U.S. population grew by 22 million people in the decade of the 1980s. On the surface, that figure looks horrific: *22 million new mouths to feed*, with the required new jobs to support them! But a critical bit of information was missing: that growth *rate* for the decade was 9.8 percent, and was the second lowest on record.

Even the multi-hundred billion dollar U.S. budget deficits of recent years don't look too awful when viewed as a percentage of gross national product (GNP). At least one paper in *Harvard Business Review* (Eisner

1993) argued that the deficit is not so horrible as people believe, and may actually stimulate the economy if the debt were incurred for the right things.

Another thing that distorts the interpretation of otherwise valid studies is the improper combining of the results of two or more studies. This mistake is the "apples and oranges fallacy." Advocates cite two or more studies on the same subject to bolster their position in a manner that suggests the studies are somehow mutually reinforcing or complementary. A conclusion drawn as a synthesis from the results of two or more non-related studies is always suspect, and may be quite meaningless.

Studies almost always have some constraints, a basic methodology, and some (hopefully not hidden) basic assumptions. If the two studies cited do not share those factors in common, then drawing any conclusion that depends on combining them is invalid. The synthesized result is meaningless unless there is some means for compensating for the fundamental differences in the studies. Unless one is counting the category "fruit," one cannot fairly compare apples and oranges. It is very difficult for a layperson to make valid comparisons in such cases, so be wary when it is done without a good reason.

Scientific Studies

Scientific studies seem to be particularly subject to abuse when presented to a lay audience. Few people outside science understand that the results of studies are always considered tentative. They also understand that very few issues can be definitively proven by a single study. The public is often confused when two or more studies seem to contradict each other. "Is vitamin-C good for the common cold or not?" "Does eating bran lower cholesterol or not?"

Science makes progress by iteratively examining issues, and then holding up new data to public, and often very brutal, examination. As a result, each new study should build on earlier work, and refine its focus to overcome objections to the earlier studies. After this is done a number of times, some good approximation of the truth should emerge. But if one takes a "snapshot" view of the research by looking at only one study at a single point in time, then an erroneous picture may be seen. Scientists understand this problem, and account for it in their thinking, but laypersons rarely know how to look at studies. All that disagreement between studies may indicate, is that scientists haven't refined their knowledge of a subject well enough to ask Nature the right question...yet.

Some General Advice

From sets developed by Cohn (1989) and Berkman (1993), supplemented by some of my own criteria, we can list several questions that should be asked of the author of any article or study:

How do you know that such-and-such is true?

Is the published data preliminary or final?

Do other experts in the field concur? Is their concurrence general or highly specific?

What is information based on?

Have the assertions been validated in a formal study or experiment?

Was the study design according to generally accepted scientific standards?

Who funded the work?

What stake does the researcher have in the outcome (reputation, $$$, promotion potential)

Who disagrees with the conclusions and why?

How sure are you of the conclusion?

Are the conclusions backed up by statistical evidence?

Have the studies been replicated by others? What were the results?

Were the results reasonably consistent from one study to the next?

Are other explanations for the observations possible?

Who else in field has seen the work? Was it peer reviewed?

What methodology was employed?

What are study's weak points?

What criticism has been received? From who?

Do you agree with the advocate's conclusions drawn from your study? Was the work fairly used?

Would competent experts in the field be surprised by the result? If so, would they find the interpretation novel and bizarre?

When evaluating a researcher, look for agendas, hidden and otherwise, the backgrounds and qualifications of the researchers, their normal job function, their self-proclaimed mission (if any), and their source of support or funding to detect possible biases in the study. Ask a real pertinent question: "Why is this person interested, and how does it affect the results?" (the word "interested" to mean "has a stake in" rather than "curious about....")

It is also relevant to know what peer recognition the researcher enjoys (or is afflicted by). It may also be pertinent to note who referred you to that researcher. Was it someone you respect? Was it a self-nominated advocate? Do they have a general reputation for reliability and integrity?

So Why Bother?

So why should you care that salt water taffy is being distributed by those trying to persuade you? Because it pulls out your dental fillings, rips off your crowns, rots your teeth, raises your blood sugar level, is hard to chew, is generally unhealthy, yet doesn't provide any form of intellectual nourishment in return for all its evils. It's icky stuff, so shun it. Better yet, stick the taffymaker to the wall with his own stuff.

One of the principal (and most telltale) signs that one is dealing with the mentally mushy is the use of a junk science approach in presenting their arguments. One hallmark of junk science is that the argument leads one to the edge of a logical chasm that is simply too broad for the healthy mind to leap. We will take a look at some of the footpaths that lead to the edge of that precipice.

Examples of junk science abound on the fringes of fields as diverse as medicine, cosmology, psychology and history. Even today, when medicine is firmly on a scientific footing, fringe elements pop up from time to time with strange new quack medical "cures" or therapeutic devices. I still see copper bracelets on the wrists of arthritis sufferers, and physicians tell me that Mexican border towns are overburdened with quacks selling spurious treatments, drugs, devices and other remedies that are not available in the USA (and for good reason—they don't pass scientific scrutiny).

Electrical or electronic devices are all-time favorites with the quack medical community. Such devices are also popular among individuals who buy them for self-treatment without medical advice. Such devices and other treatments are bad news for people who give up traditional treatments known to be effective. Because of legal restrictions on the sale of assembled units, some quack medical devices are sold in kit form and must be assembled by the end user.

"Old-time radio" nostalgia buffs may remember the superpower (250,000+ watts) Mexican "Border Blaster" AM radio stations. During the 1930s and 1940s—where "doctor" Brinkley and other "eclectic school physicians" used border radio to sell their dubious goat gland transplant "prostate" surgery for curing male impotence. They had operated in the United States, but during that period one state after another rejected "eclectic medicine" (which often required only a few months of study by mail order) and instituted licensure of physicians based on educational standards and a scientific basis of knowledge.

In the realm of the cosmos, the theories of the late Immanuel Velikovsky still gain adherents. The UFO enthusiasts are so numerous that they can hold well-attended national conferences. And of course, there are those ancient astronauts who drew the immense pictures in the highlands of Peru...never mind that the feat was duplicated by civil engineering students using crude replicas of ancient wood and metal surveying tools.

New Agers still follow the teachings of the likes of Edgar Cayce and an eccentric Englishman who claims that the Christ is here on Earth today (and since 1977 has been posing as an immigrant Pakistani taxicab driver in East London). He is said to have reentered the world from mystical Shambala Valley high in the Himalayan Mountains, using an airplane for transportation (Wow! "Christ" needs an airplane?). Benjamin Creme's "christ" is remarkably bashful...he has missed several announced coming-out dates.

The non-science world may be even more prone to junk than the scientific community. At least in science there is an established tradition that militates against the flakey. But in other fields, there is either no such tradition, or else it is very weak. One of the most interesting is the matter of conspiracy theories. While these theories seem to be the property of the far right, they are actually a lot more universal.

Pseudoscience

There are several attributes found in junk science (Casti 1989, Shermer 1997), and other poorly substantiated theories, although not all of them may be present in any particular case. Whenever you see any of these factors, however, raise the caution flag and take a deeper look at what is being offered.

Anachronistic Theories. Junk science arguments are often based on theories and beliefs that were once current, but are no longer accepted. For example, if a theory of light requires the existence of a "luminiferous aether" to propagate strange "waves," then one should remind the theorist of the Michaelson-Morley experiments more than a century ago. We still refer to the "ether" when speaking of radio propagation, but no one gives it physical credibility.

Rejected Theories. Closely related to, but different from, the use of anachronistic theories is the use of rejected theories. Where the anachronistic theory was once accepted by at least a substantial number of thinking people, and then honorably superceded by new knowledge, rejected knowledge was always on the fringe. Historian James Webb cites a fondness for rejected knowledge as the heart of what he called the Occult Revival of the nineteenth century.

Seeking Mysteries. Science sometimes stumbles across mysteries, ancient and modern, and even manages to solve a few of them from time to time. But seeking mysteries, or using mysteries in explanation of the otherwise unexplainable, is part of the method of junk science. A good example is the extremely large figures in Peru's high desert areas that are only visible by airplane. New Age enthusiasts resort to the mystery of how they were made to justify UFO and ancient astronaut theories. It was believed that no one without modern surveying equipment could lay out such straight lines over the horizon...until some civil engineering graduate students did exactly that using simple instruments of crude construction and a little bit of simple geometry.

Appeals to Myth. A myth is a story that either relates to or explains ancient historical events. Some myths are the founding story of entire cultures, and may be true, false or only partly true. Myths are ripe for wild speculation. Some people will engage in a futile exercise in circular logic by concocting a story to explain the myth, massage it into a plausible sounding theory to make it appear respectable, and then use the existence of the myth to prove the theory; round and round we go (sigh).

Weighing the Evidence (literally). Bad theories are often supported by voluminous evidence, often anecdotal, that is offered in proof of the theory. The sheer volume of the evidence is loudly and confidently proclaimed to be proof of the theory. But, they seem to forget, piling trash on top of trash does not turn a dung heap into gold. One piece of good quality evidence in support or refutation of a position is far superior to any amount of poor evidence, no matter how voluminous.

Reliance on Irrefutable Hypotheses. The hallmark of a good hypothesis is that it can be refuted. The hypothesis that an open dish of kerosene won't burn when a lighted match is tossed in is easily (and spectacularly) refutable...and probably on the first attempt.

A theory that requires critics to prove a negative ("ghosts do not exist") is of poor quality, and should not be entertained. Except, of course, on late night television where things are a little weird and most people are asleep anyway (even the awake ones).

Refusal to Revise. A characteristic of good science is the willingness to revise one's theories when defects are pointed out by critics, or when new information becomes available. Around the turn of the century Newtonian physics dominated science. But troubling observations were found that could not be explained by Newton's theories. As a result, quantum mechanics and relativity theory emerged. Physicists modified their theories to account for the new observations. Junk science advocates usually refuse to revise their theories when faced with criticism or new observations.

Whole World Against Me. Junk science is usually based on one or more fallacies, incorrect data, or "stuff" that's simply plucked from air. As a result, orthodox scientists reject the theories. If you hear someone claiming that the Establishment is against them, or covertly attempting to suppress the "true" theory (usually for money), then be wary. Unproven (or fake) drugs and spurious medical devices cannot legally be sold in the USA because they lack approval from the Food and Drug Administration (FDA). It's the "FDA conspiracy" that is supposedly keeping this wonderful advance off the market. But the truth is a little less interesting: the FDA requires scientific proof before granting approval, which is precisely what junk science cannot provide.

In the 1970s there was a cult following for medical use of an organic solvent called DMSO. One advocate told me that "...FDA is holding it off the market because it is so cheap and the big drug companies don't want competition." The big drug companies would, I suspect, be highly amused at the idea that the FDA was their protector and benefactor! Checking with the FDA yielded a less interesting truth: no one had filed an application to market the drug. It seems that DMSO was unpatentable, and in the public domain. That means that anyone can make the stuff. The drug companies were therefore not interested in spending money to do the necessary scientific trials so that other companies could benefit from their work. Yet people still bootleg supplies of DMSO and use it, despite the fact that there are dangers in DMSO that should only be evaluated by competent physicians.

Use of Anecdotes. An anecdote is a war story. Junk science often uses war stories to prove their positions. "I took *Flippinz* and my cold disappeared." The fact is that a common cold will go away in a week if you go to the doctor, or seven days if you don't. As a result, you could rub mineral water on your big toe and be "cured" of your cold if you wait long enough to make the observations. Some anecdotes are not so excusable. The person who sells a spurious "cure" for a deadly disease (e.g., AIDS, heart disease, cancer) is morally reprehensible.

The problem with anecdotes is that they are not controlled experiments, so one cannot be sure that the results are what they appear. In a controlled experiment, the scientist attempts to control or account for all variables other than the one being tested. Anecdotes might point the way to an interesting research path, but they do not establish the truth or falsehood of anything.

Abuse of Scientific Language or Making Bold Statements. Junk science advocates often resort to either overuse of scientific language (or "scientific sounding" language where the real item won't work), or the making of bold pronouncements about their position. Don't be taken in by high sounding language...look for evidence and proof.

Glorification of Heresy. There is a recognized process in science that requires new ideas to prove themselves in the crucible of brutal public criticism. The more bold the claim, the more vigorous the scrutiny. And if the new idea contradicts an old idea, the resistance is especially vigorous (Kuhn 1962). The history of science is littered with scientists whose ideas were at first rejected, only to be accepted later on. But that doesn't mean that every rejected theory is a present heresy but future orthodoxy. Again, demand evidence (and the more extraordinary the heresy, the greater the proof required).

Rumors and Urban Myth as Proof. The rumor is a powerful tool in the hands of the junk science advocate. Something that "everyone knows" is difficult to refute, if only because the holder of the belief is not open to facts. Urban myths abound. Remember the alligators in the sewers of New York City? Or how about the Kennedy assassination myths. Recall the story of the four "well dressed hobos" arrested in the train yard near the "grassy knoll." It is widely reported that these hobos disappeared from sight immediately after the assassination, implying that they were part of a hit team. It is asserted that their arrest records were suppressed...if they ever existed. However, an enterprising reporter for a Public Broadcasting System special report found the arrest record on the Dallas police blotter...a few lines above Lee Harvey Oswald's arrest record. Furthermore, they located one of the hobos, still alive, working as a dishwasher in the Pacific northwest. The "well-dressed hobo" theory was widely disseminated, widely believed and became a part of what conspiracy buffs thought was the suppressed truth.

Novel and Bizarre Interpretations. A historian was commenting on a colleague who mixed up a concoction and tried to publish it as historical fact. The colleague claimed to be the first person to see the events correctly. The historian then sarcastically replied: "John, you are the only one to ever see that...." Later on, John was boasting "Professor Smith told me I was the only one to ever see these truths...."

There is a well established maxim for interpreting evidence. It's called "Ocham's razor," after one Bishop Ocham. This maxim holds that the *most probably correct interpretation is the simplest interpretation that fully accounts for all of the known facts*. For example, when you see a light streaking across the sky at high speed, the most probable explanation is a meteor, not a UFO.

One weekend in the early 1960s there was a very large suck-out of HF signal propagation (and on ARRL Sweepstakes weekend, as I recall!). The simplest explanation was a Sudden Ionospheric Disturbance produced by a solar flare. On the 75-meter ham band, however, one chap was pontificating about space aliens blanketing our section of the cosmos with some mysterious ray that sucked up radio signals. The funny part was that a number of knowledgeable hams were stringing him along, and their responses were more humorous than the "space alien" advocate.

When wielded correctly, Ocham's razor can slice baloney exceedingly thin...so thin as to be transparent.

The Argument from Selected Instance. Junk science advocates often present what seems to be a large amount of good evidence in support of their spurious claims. On closer examination, however, it appears that they selected only that evidence which seemed to support their position, and ignored evidence that refuted it. The only competent way is to consider and explain all of the available evidence (rigorously applying Ocham's razor in the process).

There are several different approaches to the Selected Instance Fallacy. Oftentimes, there is no intent to defraud, but only a natural tendency to see confirmatory evidence ("sharpening") and not see disconfirming evidence ("leveling"). In some cases, an early run of confirming evidence is taken as strong proof of the truth of a hypothesis. As a result, subsequent data collection is either tainted by leveling, or the experiment is terminated and success is declared.

There is so much junk science and revisionist junk history around us today that one is under constant assault. The cure is to develop a skeptical mind and test everything. The skeptic has an advantage over others: a built-in high-Q BS filter.

Bibliography

Ackoff, Russell L. and Fred E. Emery (1972). *On Purposeful Systems*. Seaside, CA: Intersystems Publications (1981 reprint edition).

Adder, Jerry (1994). "The Numbers Game." *Newsweek* Lifestyle section. 25 July 1994. 56-59. New York.

Baron, Jonathan (1994). *Thinking and Deciding - 2nd Edition*. New York: Cambridge Univ. Press.

Berkman, Robert (1993), "But Is It The Whole Truth?", *Writer's Digest*, SEPT 1993, p. 42.

Brott, Armin A. (1994). "Battered-Truth Syndrome: Hyped Stats on Wife Abuse Only Worsen The Problem." *The Washington Post*, Outlook Section, 31 July 1994, pp. C1-C2.

Casti, John L (1989), *Paradigms Lost: Images of Man in the Mirror of Science*. New York: William Morrow & Co., Inc.

Cohn, Victor, 1989, *News & Numbers*, Ames, Iowa: Iowa State Univ. Press.

Crook, Kenneth B.M. (1961), "Suggestions for Teaching the Scientific Method," *American Biology Teacher*, March 1961.Copyedited manuscript version supplied to the author by Dr. Crooks' family.

Crossen, Cynthia (1994). *Tainted Truth: The Manipulation of Numbers in America*. New York: Simon & Schuster.

Eisner, Robert (1989), "Sense and Nonsense About Budget Deficits," *Harvard Business Review*, Vol. 7, No. 3, May-June 1993, p. 99.

Englebretsen, George (1995). "Postmodernism and New Age Unreason." *Skeptical Inquirer* pp. 52-53. May/June, Vol. 19, No. 3. Amherst, NY.

Flew, Antony (1977). *Thinking Straight*. Buffalo, NY: Prometheus Books.

Gilovitch, Thomas (1991). *How We Know What Isn't So: The Fallibility of Human Reason in Everyday Life*. New York: Macmillan/Free Press.

Jones, Morgan D. (1995). *The Thinker's Toolkit: Fourteen Skills for Making Smarter Decisions in Business and in Life*. New York: Random House.

Kahneman Daniel, Paul Slovic and Amos Tversky (1982), *Judgement Under Uncertainty: Heuristics and Biases*, New York: Cambridge Univ. Press.

Kinsey, Alfred C., Wardell B. Pomeroy and Clyde E. Martin (1948), *Sexual Behavior in the Human Male*. Philadelphia: W.B. Saunders.

Kuhn, Thomas (1962), *The Structure of Scientific Revolutions*. Chicago: Univ. of Chicago Press.

Lewis, C.I. (1929). *Mind and the World Order*. New York: Charles Scribner & Sons. Reprint edition: Dover Publications.

McKean, Kevin (1985), "Of Two Minds: Selling the Right Brain," *Discover*, April 1985, pp. 30-38.

McLaughlin, J.P. (1995). "On Leaping and Looking and Critical Thinking." *Skeptical Inquirer*, pp. 6-7. May/June, Vol. 19, No. 3. Amherst, NY.

Morin, Richard (1994a). "How to Lie With Statistics: Adultery."*The Washington Post*, 6 March 1994, p. C5.

Morin, Richard (1994b). "Racism on the Left and Right." *The Washington Post*, 6 March 1994, p. C5.

Neustadt, Richard E. and Ernest R. May (1986). *Thinking in Time: The Uses of History for Decision Makers*. New York: The Free Press.

Pomeroy, Wardell B. (1972). *Dr. Kinsey and the Institute for Sexual Research*. pp. 292-293. Cited in Reisman 1990.

Reisman, Judith A. and Edward W. Eichel (1990). *Kinsey, Sex and Fraud*. Lafayette, LA: Huntington House.

Ruchlis, Hy (1990), *Clear Thinking: A Practical Introduction*, New York: Prometheus Books.

Schick, Theodore Jr. and Lewis Vaughn (1995). *How to Think About Weird Things: Critical Thinking for a New Age*. Mountain View, CA: Mayfield Publishing Co.

Selltiz, Claire, Lawrence S. Wrightsman, and Stuart W. Cook (1976), *Research Methods in Social Relations — 3rd Edition*, Holt, Rinehart and Winston (New York, 1976).

Shermer, Michael (1997). *Why People Believe Wierd Things*. New York: W.H. Freeman Co.

Sommers, Christina Hoff (1994). *Who Stole Feminism?: How Women Have Betrayed Women*. New York: Simon & Schuster.

Webster's New Collegiate Dictionary (1979), Springfield, MA: G.& C. Merriam Company.

Index

C

D

PROMPT®
PUBLICATIONS

Electronic Circuit Guidebook, Volume 1: Sensors
Joseph Carr

Many texts discuss instrumentation and computer interfacing. A topic usually lacking in these texts, however; is covered in *Electronic Circuit Guidebook, Volume 1: Sensors* - namely, the analog interface.

In this volume you will find information you need about typical sensors, along with much more information about analog sensor circuitry. Amplifier circuits are well covered, along with differential amplifiers, analog signal processing circuits, and more.

Topics covered include: Noise, Sensors and Instruments, Instrument Design, Resistive Sensors, Electro-Optical Sensors, Magnetic Sensors, Interfaces, Building the Analog Subsystem, Analog Amps, Inverting and Noninverting Amps, Differential Amps, Analog Signal Processing Circuits, Removing 60 Hz Noise, Shield and Grounding Methods, A/D-D/A Converters, Sensor Resolution Improvement Techniques, DC Supply Circuits, AND MORE.

Electronic Circuit Guidebook, Volume 2: IC Timers
Joseph Carr

Part I of this book demonstrates the theory of how various timers work. This is done by way of an introduction to resistor-capacitor circuits, and in-depth chapters on TTL and CMOS digital IC timers, the LM-555 and other ICs, operational amplifier timer circuits, retriggerable timers and long duration timers. The simplified equations and detailed graphics are perfect for both technicians and hobbyists.

Part II presents a variety of different circuits and projects, some standalone, and others for incorporation into other circuits. Examples included: touchplate trigger, missing pulse detector, analog audio frequency meter, one-second timer/flasher, relay and optoisolator drivers, two-phase digital clock, 100kHz crystal calibrator and many more. *Electronic Circuit Guidebook, Volume 2: IC Timers* will teach you enough to be able to rework and modify the circuits covered here, and also design a few of your own.

Electronics Technology
339 pages • paperback • 7-3/8 x 9-1/4"
ISBN: 0-7906-1098-1 • Sams: 61098
$29.95

Electronics Technology
239 pages • paperback • 7-3/8 x 9-1/4"
ISBN: 0-7906-1106-6 • Sams: 61106
$29.95

**To order your copy today or locate your nearest Prompt®
Publications distributor : 1-800-428-7267 or www.hwsams.com**
Prices subject to change.

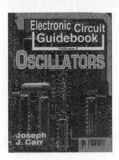

Electronic Circuit Guidebook, Volume 5: Digital Electronics
Joseph Carr

Electronic Circuit Guidebook, Volume 5: Digital Electronics is about the basics of digital electronics. It is not a computer book, but rather a book about the fundamental circuits that make up not only computers, but all digital products. It joins other books in the *Electronic Circuit Guidebook* series by providing you with encapsulated practical information about specific areas of electronics technology.

Volume 5 looks at logic gates, flip-flops, counters and the ways they are combined to make practical circuits. It also covers the following: Introduction to Digital Electronics; Number Systems; Digital Codes; Gates: The Basic Building Block; Unclocked Flip-Flops; Clocked Flip-Flops; Arithmetic Circuits; Counters; Display and Decoder Devices; Registers; Data Multiplexers and Selectors; Monostable Multivibrators; Clocks and Astable Multivibrators; Data Converters; Microcontrollers; AND MORE!!

Electronics Technology
246 pages • paperback • 7-3/8 x 9-1/4"
ISBN: 0-7906-1129-5 • Sams 61129
$29.95

Electronic Circuit Guidebook, Volume 6: Oscillators
Joseph Carr

Electronic oscillator circuits are used to generate repetitive or periodic waveforms. These waveforms may be sine waves, square waves, triangle waves, sawtooth waves or pulses, depending on the needs of the circuit being developed. It is relatively easy to make a circuit oscillate, as anyone who has tried to build certain types of amplifiers will (to their misery) testify. But to make the oscillator produce the correct amplitude, frequency and waveshape takes a little doing. In this book, you will learn how to make the "doing" a little less traumatic.

Electronic Circuit Guidebook, Volume 6: Oscillators covers : Resonant Tuned RF Circuits; RC Timing Networks; Effect of RLC Networks on Waveforms; Monostable and Astable Multivibrators; Integrators and Differentiators; Triangle and Sawtooth Generators; many types of Oscillators;The 555 Integrated Circuit Timer; and much more.

Electronics Technology
256 pages • 7-3/8 x 9-1/4"
ISBN: 0-7906-1185-6 • Sams 61185
$34.95

To order your copy today or locate your nearest Prompt®
Publications distributor : 1-800-428-7267 or www.hwsams.com
Prices subject to change.

5+M

AGREEMENT

READ THIS AGREEMENT BEFORE OPENING THE SOFTWARE PACKAGE

BY OPENING THE SEALED PACKAGE YOU ACCEPT AND AGREE TO THE FOLLOWING TERMS AND CONDITIONS PRINTED BELOW. IF YOU DO NOT AGREE, DO NOT OPEN THE PACKAGE AND RETURN THE SEALED PACKAGE AND ALL MATERIALS YOU RECEIVED TO HOWARD W. SAMS & COMPANY, 2647 WATERFRONT PARKWAY EAST DRIVE SUITE 100 INDIANAPOLIS, IN 46214-2041 (HEREINAFTER "LICENSOR") WITHIN 30 DAYS OF RECEIPT ALONG WITH PROOF OF PAYMENT.

Licensor retains the ownership of this copy and any subsequent copies of the Software. This copy is licensed to you for use under the following conditions:

Permitted Uses. You may: use the Software on any supported computer configuration, provided the Software is sued on only one such computer and by one user at a time; permanently transfer the Software and its documentation to another user, provided you retain no copies and the recipient agrees to the terms of this Agreement.

Prohibited Uses. You may not: transfer, distribute, rent, sublicense, or lease the Software or documentation, except as provided herein; alter, modify, or adapt the Software or documentation, or portions thereof including, but not limited to, translation, decompiling, disassembling, or creating derivative works; make copies of the documentation, the Software, or portions thereof; export the Software.

LIMITED WARRANTY, DISCLAIMER OF WARRANTY

Licensor warrants that the optical media on which the Software is distributed is free from defects in materials and workmanship. Licensor will replace defective media at no charge, provided you return the defective media with dated proof of payment to Licensor within ninety (90) days of the date of receipt. This is your sole and exclusive remedy for any breach of warranty. EXCEPT AS SPECIFICALLY PROVIDED ABOVE, THE SOFT-WARE IS PROVIDED ON AN "AS IS" BASIS. LICENSOR, THE AUTHOR, THE SOFTWARE DEVELOPERS, PROMPT PUBLICATIONS, HOWARD W. SAMS & COMPANY, AND BELL ATLANTIC MAKE NO WARRANTY OR REPRESENTA-TION, EITHER EXPRESS OR IMPLIED, WITH RESPECT TO THE SOFTWARE, INCLUDING ITS QUALITY, ACCURACY, PERFORMANCE, MERCHANTABILITY, OR FITNESS FOR A PARTICULAR PURPOSE. IN NO EVENT WILL LICENSOR, THE AUTHOR, THE SOFTWARE DEVELOPERS, PROMPT PUBLICATIONS, HOWARD W. SAMS & COMPANY, AND BELL ATLANTIC BE LIABLE FOR DIRECT, INDIRECT, SPECIAL, INCIDENTAL, OR CONSEQUENTIAL DAMAGES (IN-CLUDING BUT IS NOT LIMITED TO, INTERRUPTION OF SERVICE, LOSS OF DATA, LOSS OF CLASSROOM TIME, LOSS OF CONSULTING TIME) OR LOST PROFITS ARISING OUT OF THE USE OR INABILITY TO USE THE SOFT-WARE OR DOCUMENTATION, EVEN IF ADVISED OF THE POSSIBILITY OF SUCH DAMAGES. IN NO CASE SHALL LIABILITY EXCEED THE AMOUNT OF THE FEE PAID. THE WARRANTY AND REMEDIES SET FORTH ABOVE ARE EXCLUSIVE AND IN LIEU OF ALL OTHERS, ORAL OR WRITTEN, EXPRESSED OR IMPLIED. Some states do not allow the exclusion or limitation of implied warranties or limitation of liability for incidental or consequential damages, so that the above limitation or exclusion may not apply to you.

GENERAL:

Licensor retains all rights, not expressly granted herein. This Software is copyrighted; nothing in this Agree-ment constitutes a waiver of Licensor's rights under United States copyright law. This License is nonexclu-sive. This License and your right to use the Software automatically terminate without notice from Licensor if you fail to Comply with any provision of this Agreement. This Agreement is governed by the laws of the State of Indiana.